UTAH'S HIDDEN TREASURE

OUTLAW LOOT IN EVERY COUNTY

UTAH'S HIDDEN TREASURE

OUTLAW LOOT IN EVERY COUNTY

STEPHAN B. SHAFFER

AN IMPRINT OF CEDAR FORT, INC.
SPRINGVILLE. UTAH

ISBN 13: 978-1-4621-2056-7

Published by Plain Sight Publishing, an imprint of Cedar Fort, Inc., 2373 W. 700 S., Springville, UT 84663
Distributed by Cedar Fort, Inc., www.cedarfort.com

LIBRARY OF CONGRESS CATALOGING-IN-PUBLICATION DATA

Names: Shaffer, Stephan B., author.
Title: Utah's hidden treasure : outlaw loot in every county / by Stephan B. Shaffer.
Description: Springville, UT : CFI, an imprint of Cedar Fort, Inc., [2017] | Includes bibliographical references and index.
Identifiers: LCCN 2017013428 (print) | LCCN 2017018662 (ebook) | ISBN 9781462128099 () | ISBN 9781462120567 (pbk. : alk. paper)
Subjects: LCSH: Treasure troves--Utah--History. | Mines and mineral resources--Utah--History. | Outlaws--Utah--History. | Utah--Antiquities.
Classification: LCC F828 (ebook) | LCC F828 .S53 2017 (print) | DDC 979.2/01--dc23
LC record available at https://lccn.loc.gov/2017013428

Cover design by M. Shaun McMurdie

Typeset by Chelsea Holdaway

Printed in the United States of America

10 9 8 7 6 5 4 3 2 1

Printed on acid-free paper

I dedicate this book to those who have given me information regarding outlaw loot, mines, and events that have shaped Utah's history.

Thanks to my wife, Bonnie, for standing by me and supporting me in my quests for truth.

I want to thank my friends who are more than just friends but who help discover historical events that have been overlooked or forgotten.

Also by Stephan B. Shaffer

La Mina del Yutas

Nachi: Man of Justice, Son of Warriors

Of Men and Gold

Out of the Dust: Utah's Lost Mines and Treasures

Treasures of the Ancients: Recent Discoveries of Ancient Writings in North America

Voices of the Ancients

Contents

Old Spanish mine in Soap Stone Basin, Uinta Mountains

Introduction

Almost everyone has dreamed, at one time or another, of finding a lost gold mine. If my dreams are different from the dreams of others, it is because I have always hoped that most of my finds would have historical value. I have found mines (some good ones too), but the riches in them brought me more grief than joy.

I felt that a tour through the state of Utah, with a tidbit of information about other areas close by, would be a good way of getting the juices flowing in the reader's veins, inspiring them to get up off the couch, put down the TV remote, hide the iPod or whatever device that keeps them inside, and venture out amongst the hills and valleys in search of that lost mine, treasure, or outlaw loot!

Having spent many years (mostly in my early years) hunting such things, I learned a lot about many places, and I accumulated dozens of stories and maps. Are they all true? Maybe not, but that's part of the excitement! It seems that most stories have been somewhat embellished, and many maps have been drawn to scam some unknowing greenhorn. But there are good maps and good stories, and it's an adventure to figure out what is and what isn't true.

In this book, I won't sell you a fake map or give you a false story on purpose. If I feel that it is a bit on the shady side, I'll tell you, and if a map I show you is not accurate, or if I think it is not accurate, I'll tell you. So, take your kids and your spouse and go and have fun. That's what it is all about anyway.

Steve Shaffer

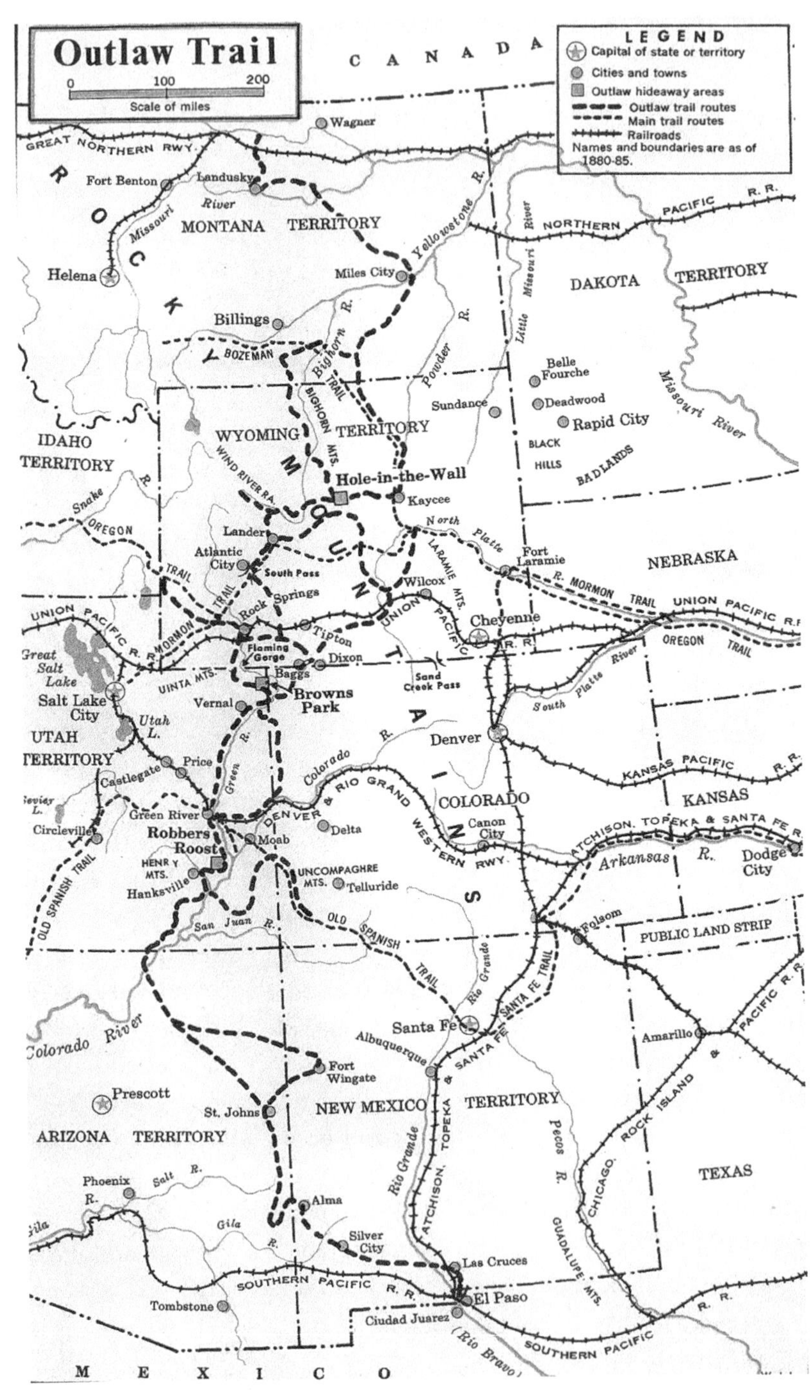

The old outlaw trails

Beaver County

"Beaver County was created in 1856; it is a parallelogram extending from the crest of the Tushars on the east to Nevada on the west. Mineral, San Francisco, Wah Wah, and other Ranges [Indian Peak and part of the White Rock Mountains on the border with Nevada] . . . are noted for their minerals. One of Utah's few sulfur mines is in the northeast corner [of the county]."[1] Beaver is the county seat, and one of the noted events was the trial of John D. Lee, the mastermind of the Mountain Meadows Massacre.[2]

LOST MINE IN THE MINERAL MOUNTAINS

There are countless stories of lost loot, Spanish mines and treasures, and lost mines in Beaver County, such as the story of a lost Spanish mine located in the Mineral Mountains just west of Beaver. The mine was discovered in 1858 by some prospectors looking for minerals. That day, they found what turned out to be a very rich lode of galena ore. This mine was called "The Lost Spanish Mine" because of the old rusty tools found at the site. The prospectors could tell the mine hadn't been worked for many years, and the only people in that part of the country besides the local Indians were the Spaniards. When the ore was taken to Beaver for smelting, it was soon discovered that their "galena" was more silver than lead! A town sprung up soon after and was named Lincoln, after President Lincoln. Before long, the area was full of prospectors and miners digging holes all through the mountains in search for their share of the rich ore hidden beneath the rocky crags of the Mineral Mountains. Many hid their discoveries from the prying eyes of would-be thieves only to lose them themselves.

TREASURE IN THE WAH WAH MOUNTAINS

Many mines sprang up along the Wah Wah Range, and the mountain was crawling with men looking for mineral wealth. These mountains are well-known for their supply of red beryl and smoky topaz crystals. If the right piece is found, red beryl can be worth more than gold.[3] Why this happens is anyone's guess! In 1980, a large 38-carat red beryl was discovered on the west flank of the Wah Wahs.[4]

HIDDEN CACHES AT FRISCO

There are stories of hidden caches at the old ghost town ruins of Frisco. In its heyday, there was a murder a night and the streets were lined with saloons, eateries, gambling dens, and brothels. Food and drink cost so much that a normal citizen could not afford to purchase much in the town.[5]

Today, relic hunters still go there with their fancy metal detectors in the hopes of finding one of the many caches said to be hidden in and around the old town. Once in a while, we hear of someone making a discovery that brings the adventurous bug out in all of us, driving us to get back to those hills in search for a lost cache!

THE LOST GARNET MINE

Out on the southern end of the White Rock Mountains, not far from the border, there is a hidden garnet mine. The mine is said to be of Spanish origin, and Spanish artifacts were discovered in the area with a pack of garnets. One prospector claimed that he found the mine but lost track of it after not being in the area for a number of years. Even though he looked for quite some time, he could never find it again.

According to this man, the garnets he found were the size of a man's little fingernail and were brightly colored. When asked why he thought they were garnets, he said that he had two of them looked at by a jeweler in Salt Lake City, Utah.

If you want to go look for the Lost Garnet Mine, you should start at Paradise Canyon, twenty miles north of Modena. Somewhere close to that canyon lies the Lost Garnet Mine.

THE CAVE MINE

The story of the Cave Mine is known by many folks in and around Beaver. From sources gleaned from these folks, we know that John Bradshaw was the man that found the Cave Mine. Born in England, Bradshaw came to America during the 1860s. What brought him to Utah is unknown, but folks seem to think he was interested in mineralogy and wanted to find either gold or silver in the surrounding mountains.

"One night, he had a dream of a white mule standing on a ridge. He dreamed that he went to the ridge north of Minersville and arrived at a cave that had never seen by anyone before. He [went] down inside the cave and found nuggets of gold and there were many rats' nests there."[6] The dream so impressed him that he was sure he could find this cave, so he gathered up his gear and headed out. He walked many miles until he reached an area of thick trees; he began to feel that he might get lost if he went into them too far. However, he felt sure that if he did go on, he would find the cave in his dream.

Bradshaw kept on going. He tied strings of cloth to trees so he could find his way out of the forest. To his surprise, he found the cave just like in his dream; even the rats were the same. He found his way out by following the strips of cloth, and he headed back home. Once there, he persuaded someone to go back with him to help lay claim to the old cave. The cave was located in the Bradshaw Mining District and turned out to be a very good producer of gold, silver, and lead. "Bradshaw died in Minersville, never having realized much out of it as, according to reports, he was swindled so much in this venture."[7]

However, the Cave Mine turned out to be a huge success and claimed to have millions in gold and silver. Every day, two to six mule teams hauled ore from the Cave Mine to the town of Frisco. Today, the mine is closed because of the decline in minable ore.

There is an old legend that says a curse was placed on the mine, stating that it will not produce any more gold and silver until a member of the Bradshaw family owns the mine again.

NOTES

1. Rufus Wood Leigh, *Five Hundred Utah Place Names: Their Origin and Significance* (Deseret News Press: Salt Lake City, 1961), 5.
2. Ibid.
3. "What Gemstone Is Found in Utah that Is Rarer than Diamond and More Valuable than Gold?," Carl Ege, *Utah Geological Survey*, accessed April 3, 2017, geology.utah.gov/map-pub/survey-notes/glad-you-asked/what-utah-gemstone-is-rarer-than-diamond/.
4. Interview with author, wishes to remain anonymous.
5. "Frisco Ghost Town Site, UT," *Southwest Backcountry: Exploring the West, Discovering the Past*, accessed April 3, 2017, southwestbackcountry.wordpress.com/category/ruins/page/5/.
6. "John White Bradshaw and the Discovery of Cave Mine—Minersville," *Facebook*, posted by Gold Rush Expeditions, Inc., April 14, 2014, accessed April 3, 2017, www.facebook.com/pg/GoldRushExpedinc/posts/.
7. Ibid.

Box Elder County

"Box Elder County was organized in 1856; it embraces a very large area of the northwest section of the State, extending from the west spur of the Wasatch—from Brigham City to the Idaho boundary—westward to Nevada. It includes much of the Great Salt Lake and the north part of the Great Salt Lake Desert; on the east are the lower course and deltas of the Bear River, the Malad River Valley, and Promontory Mountains."[1]

TREASURES LEFT BY DESERT TRAVELERS

One of the crossings taken by westward-bound travelers and pioneers would at times skirt the north end of the Great Salt Lake and then head toward the Great Salt Lake Desert. This journey was very treacherous to say the least. The Bartleson Wagon Train took this northern route and suffered untold hardships. Their oxen fell from pure exhaustion, causing the drivers to take what meager possessions they could carry and finish the journey on foot.

During the California Gold Rush, overeager prospectors and other travelers foolishly took this supposed shortcut to the gold fields. Again, countless wagons, household goods, equipment, and supplies were abandoned along the trail. Extra guns and bullets were wrapped in waterproof tents and hidden in the rocks. To date, these guns and other supplies have never been found.

The Bidwell Wagon Train crossed as well, but Bidwell, fearing for his life, buried a box of gold and silver coins along the route. He was killed by Indians along the Humboldt River. His cache was never recovered.

> In 1846, a well-equipped wagon train led by George Donner and James Reed left Illinois for California, all the members of the party loaded with all of their worldly possessions. One of the wagons was known to be carrying $15,000 in gold coins. When the train reached Fort Bridger, the party split, some wanting to take the well-known Fort Hall Trail, the remainder, led by Donner, took the little-known Hastings Cut-Off Trail. By the time they reached the Great Salt Lake Desert, the animals and travelers were already weary and the long waterless stretch was devastating and the abandonment of nonessentials began. Donner, himself, buried his own chest of gold coins after the fifth day of travel over the desert, at a campsite somewhere in the area of Floating Island within sight of Pilot Peak to the west, planning was to return to recover the treasure. But, Donner and 43 others of the party never reached California, perishing enroute. This entire route was littered with quantities of artifacts and relics and has a number of known treasure caches.[2]

There have been researchers that claim a cache left by the Donner party is at a spring north of Pilot Peak.

LOST IRON DOOR CAVE

Many people who search for lost loot have also heard of the Lost Iron Door Cave. There have been dozens of stories written about it. Some are laced with fanciful words and embellished accounts while others barely brush by with little fanfare.

There are accounts that say the iron door came from a bank vault that was robbed during a payroll, but who really knows for sure? The legend of the door has one name attached to it: Jim Polk. Polk headed a gang of thieves that robbed banks or anything he could find to rob. Legends say that the gang kept their loot behind an iron door that was affixed to the face of a cave.

The last time Polk and his gang tried to pull off a robbery, they were in Malad, Idaho, and they were recognized by the local constable. An account, discovered in Malad by the late George Thompson, claims that the ensuing gunfight took the lives of all the gang members, while another account claims that one man was left alive by the name of Dakota, but he soon died of his wounds. It was said that Dakota confessed just prior to his passing; he even gave up the location of the Iron Door Cave. Some believed him, but most thought that he was delirious due to the amount of pain he was in.

Over the years, attempts have been made to locate this mysterious Iron Door Cave, but only one account suggests that it was ever seen after the deaths of the outlaws. A cowboy from one of the ranches claimed he saw an iron door that appeared to be hanging on the face of a cave, but he couldn't recall exactly where it was. Many a prospector and treasure seeker have looked for the Iron Door Cave. The Iron Door Cave is somewhere in the hills, along the rocky ledges, on either side of Hansel Valley. Some say that the cave holds thousands of dollars in gold, silver, jewelry, and more. What a prize it would be!

OUTLAW CACHE NEAR CORINNE

"Outlaws robbed a Colorado & Southern railroad north of Colorado Springs that netted $105,000 in cash and an estimated $40,000 in diamonds, watches, and other jewelry taken from the passengers. The outlaws made a successful getaway and made their way to the area of Bear River . . . north of Corinne, where they buried the loot. The outlaws went into town where they were discovered and captured and the treasure cache was never recovered. It is said that a few individual watches and rings have been found in the Bear River area."[3]

In that same area, there is a cache of over $60,000 from other bank and train robberies, hidden by other gangs that never made it back to recover it.

During its heyday, Corinne wasn't the little lazy town it is today. It once had a population of one thousand, including a large contingent of Chinese laborers who worked for the railroad. It was a tough place to live then. There were twenty or so saloons, gambling houses, and dance halls, not to mention houses of prostitution. But it finally had its downfall in 1872 when a serious diphtheria epidemic wiped out hundreds and made others leave in droves.[4]

HOLDUP AT GROUSE CREEK

Many stories have been written about a stagecoach holdup along an old road in the Grouse Creek Mountains. The story tells of a stage driver, Harold Roost, and his only passenger heading out on this tragic trip. Onboard the stagecoach, beside the driver and his passenger, was a heavy, locked chest containing gold and silver. Their trip north, toward Burley, Idaho, was interrupted by five outlaws with other ideas for where that locked chest was to be unloaded. The passenger was killed instantly by the blast of a sawed-off shotgun, and Roost was badly wounded and left for dead. As he lay still, facedown in the dirt, he could hear the voices of the outlaws screeching in delight at the cache of gold and silver coins!

Roost was able to make his way north until he fainted from loss of blood. He was soon discovered by a local rancher who took him to Burley where he was treated for his wounds. After the law interviewed Roost, a posse was called up and began searching for the murderers and robbers. By pure chance, the posse caught up to the outlaws near the Idaho border in some of the roughest country around. While the outlaws were watering their horses, the posse made its move; but, to their surprise, one of the outlaws saw them and opened fire on the posse. During the gunfight, all of the outlaws, save one, were killed. The last living outlaw played injured and called for help. Members of the posse went to his aid, but as soon as they got close, the outlaw released a blast from his shotgun, killing one of the posse members and wounding the others. No sooner had he got that shot off than he was killed instantly by a hail of bullets from the still-standing posse members.

Soon, an intense search of the area was made in an attempt to locate the chest of coins, but it was never found. Searchers went back to the site of the ambush many times looking for the chest, but they came away disheartened. It would be years later that a modern treasure hunter with a metal detector would find the remains of what appeared to be an old strongbox, but there were no coins near it.[5]

Somewhere in the northern part of the Grouse Creek Mountains, near the Idaho border, lies a cache of gold and silver coins.

NOTES

1. Rufus Wood Leigh, *Five Hundred Utah Place Names: Their Origin and Significance* (Deseret News Press: Salt Lake City, 1961), 8.
2. "Donner Party Treasure . . . Is It Still There?," H. Charles, *Treasure Illustrated*, forum, March, 10, 2015, accessed April 4, 2017, treasureillustrated.com/Forum/showthread.php?tid=9413.
3. "Train & Stage Robberies," Floyd Mann, *The Treasures of Utah*, forum, accessed April 4, 2017, thetreasuresofutah.yuku.com/topic/951 /TRAIN-amp-STAGE-ROBBERIES#.WOPr_xLyuis.
4. *Wikipedia*, s.v. "Corinne, Utah," last modified March 18, 2017, en.wikipedia.org/wiki/Corinne,_Utah#History. See also Federal Writer's Project, *The WPA Guide to Utah: The Beehive State* (Trinity University Press, 2013); "Collinston History: Corinne," *Box Elder County*, accessed April 4, 2017, www.boxeldercounty.org/index.htm.
5. Thomas P. Terry, *United States Treasure Atlas* 9, (Specialty Publishing, 1985). Used by permission.

Cache County

"Cache Valley is a rich agricultural valley in north Utah between the main Wasatch Range on its east and the Wasatch front spur—Wellsville Mountain—on its west. Bear River traverses the northwestern corner. The valley was one of concentration and rendezvous for beaver trappers, 1820–1839. The French-Canadians and other mountain men concealed their peltries and stores in suitable sites in the valley. The word *cache* derives from the French verb *cacher*, 'to hide.' The noun *cache* was widely current during the trapper era for both the hiding place and the things hidden. Thus, Cache Valley came to be known by this name because of the *caches* in it. Cache County, organized in 1856, comprising the Valley and its watershed, rightfully took the well-established name."[1]

TREASURES IN TONY GROVE

Cache County has its fair share of lost loot and buried treasures. There have been many stories circulating over the years about lost treasures in the Tony Grove area. Residents talk about how two outlaws, fleeing from a posse, buried gold coins near the lake in Tony Grove. This cache has never been found.[2]

Another cache with over $75,000 in gold coins, was reportedly lost or buried in Logan Canyon near its mouth. It too has never been recovered.

In the late 1970s, two prospectors found a dozen gold coins near Tony Grove. In the 1980s, a fisherman found a $20 gold piece in the Logan River.

MILLVILLE CANYON MINE

Two boys chasing cows in Millville Canyon stumbled into an old mine. After playing in the deep tunnel, they returned to finish pushing the cows down the canyon. When one of the boys took off his coveralls off for his mother to wash them, she found a large gold nugget in the cuff! The boy and his father went back up the canyon to where the boys claimed the mine was, but they never found it.

GOLD AT PORCUPINE RESERVOIR

In 1977, a prospector was metal-detecting in Paradise Dry Canyon (south of Logan and just north of Porcupine Reservoir) when all of a sudden his metal detector alerted him of something buried. Digging with his spade, he uncovered two $20 gold pieces. Further detecting the area, he eventually uncovered seven more $20 gold pieces. The treasure hunter refused to be identified, fearing retribution from officials.

CACHE VALLEY RENDEZVOUS

In our history books, we can read about an important historical event known as the Mountain Man Rendezvous of Cache Valley; it took place in the mid-1800s. Unfortunately, these rendezvous encampments only lasted about fifteen years, but hundreds of thousands of furs, hides, bottles of whiskey, women, horses, scalps, and weapons of war were traded for, and many were fought for to the point of death!

In 1826, a major rendezvous took place at Cache Valley. Jedediah Smith, a well-known trapper and explorer, had wintered on the Republican River in northwestern Kansas from 1825 to 1826. He joined General William Ashley and the supply train on the trail from St. Louis to the Cache Valley rendezvous in the spring.[3] A new batch of enterprising young men had come west with the supply train, and it was largely from the ranks of these new men (whom he had been able to observe during the long weeks on the trail), that Smith, in consultation with his partners, selected those who would make up the southwest expedition.

Unfortunately, very little is known about the members of the party outside of Smith himself except that they seem to have been a cross section of frontier America—in other words, they cut straight across to their intended goal without wavering. Harrison G. Rogers was named clerk and second in command.[4]

The party arrived at Cache Valley to find it full of traders, trappers, Indians, and more. In those days, a trapper could expect to get between $4 and $6 for each of his pelts, depending on quality. At top price ($6), trappers had to bring in 170 dressed beaver skins to get a credit of $1,000. Everything sold at these rendezvous was largely inflated, sometimes as much as 2,000 percent over the St. Louis prices.[5]

THE BLACKSMITH FORK RIVER CACHE

It was during the 1826 rendezvous that a party of independent trappers cached a large store of furs along the banks of the Blacksmith Fork River near present day Hyrum, Utah. They were afraid the pelts would be stolen from them by the larger fur companies in attendance. They wrapped their furs in waterproof, airtight bundles and buried them. According to one source, it was a large and impressive cache.[6] A fight broke out amongst some hot-headed Indians and the small independents. It is thought that the trappers were either killed or wounded and were only able to tell of the buried cache. According to source material of that time, the cache was estimated to be worth $150,000 in 1826. As far as anyone knows, it is still buried somewhere along the river near Hyrum, Utah.[7]

GOLD AT SPRING CREEK HOLLOW

It was reported that in 1950, a young ranch hand named Gary Hunsaker found several gravel-sized gold nuggets while he was out rounding up strays between Mantua and Paradise, somewhere along SpringCreek Hollow. After picking up as many pieces as he could readily see and putting them in his saddlebags, he erected a small rock monument so he would be able to find the spot again. Winter set in quickly that year, so Hunsaker did not get back to his "glory hole." The following spring, he headed back into Spring Creek Hollow. To his surprise, he was unable to locate the little rock monument or any of the other landmarks he had seen.

Hunsaker told this story to a few other ranch hands, and a few years later, another young lad came to him with three large gold nuggets that he had dug from a rotten quartz seam in the near vicinity. Hunsaker asked the boy to draw him a map so he could go look for the gold. The young man drew up a map. Even with the map, Hunsaker wasn't able to make the discovery of the rotten quartz seam or find any more gold nuggets.

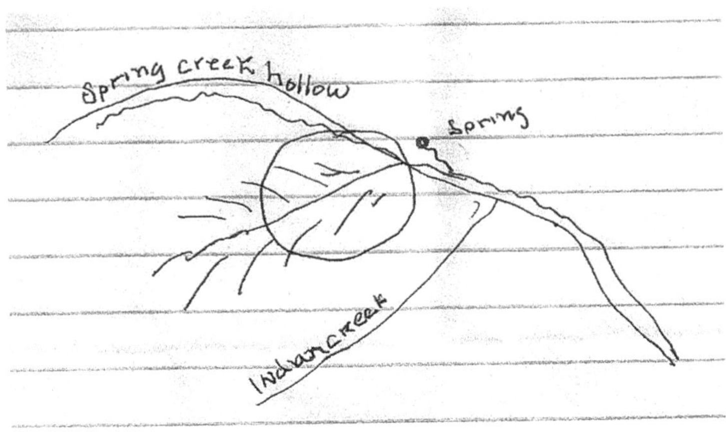

Gary Hunsaker map

NOTES

1. Rufus Wood Leigh, *Five Hundred Utah Place Names: Their Origin and Significance* (Deseret News Press: Salt Lake City, 1961), 10–11; italics in original.
2. "State Treasures: Utah," *Lost Treasures USA*, accessed April 4, 2017, www.losttreasureusa.com/StateTreasures.htm.
3. Jay H. Buckley and Jeffrey D. Nokes, *Explorers of the American West: Mapping the World through Primary Documents* (Santa Barbara, CA: ABC -CLIO, 2016), 215.
4. William Henry Ashley, Jedediah Strong Smith, Harrison G. Rogers, *The Ashley-Smith Explorations and Discovery of a Central Route to the Pacific 1822–1829,* edited by Harrison Clifford Dale (Cleveland: Arthur H. Clark Company), 194.
5. Alson Jesse Smith, *Men Against the Mountains: Jedediah Smith and the South West Expedition of 1826–1829* (John Day Company, 1965).
6. Ibid.
7. Ibid.

Carbon County

"Carbon County is a parallelogram extending from the crest of the Wasatch Plateau eastward to the Green River in east-central Utah. Carbon County was organized in 1894. The name Carbon is in reference to the immense deposits of coal and hydrocarbon shale within the county."[1]

Outlaws, Spaniards, Indians, prospectors, and prostitutes were the norm during the nineteenth century in Price, but there was more. Stage and freight lines moved through the town to places like Wellington and up to Nine Mile Canyon, past Nutter Ranch, and on to Myton, Roosevelt, Fort Duchesne, and Vernal. These trips were long and hard by themselves, but with outlaws, marauding Indians, or the occasional flash flood, these trips could be almost unbearable.

MATT WARNER: THE OUTLAW TURNED DEPUTY SHERIFF

One of the county's most famous citizens was Willard Erastus Christianson, also known as Matt Warner, Ras Lewis, and The Mormon Kid (1864–1938).

> Both an outlaw and a lawman, Christianson was born in Ephraim, Utah in 1864 to a Swedish father and a German mother who had come to Utah as converts to the Mormon Church. Though his start was good, he got into

Matt Warner in prison

a fight when he was 14 years old, and fearing he had beaten the other boy to death, he ran away. He soon joined up with a band of rustlers to begin his life as an outlaw. It was at this time that he began going by the name of Matt Warner. Somewhere along the line, he got married to a girl named Rose Morgan and the two ran a cattle ranch in Big Bend, Washington before returning with his wife and a daughter to Utah.

He then hooked up with his brother-in-law, outlaw Tom McCarty. In no time, Warner was robbing banks and trains with the likes of Elza Lay and Butch Cassidy. He then got into a shoot out that earned him five years in the Utah State Prison. Though he received an early release for good behavior, his wife died during his incarceration.

After his release, he remarried and settled in Carbon County, Utah. Warner ran for public office under his real name, Willard Erastus Christianson, and lost. He then had his name officially changed to Matt Warner, the name most people knew him by, and was elected justice of the peace and then served as a deputy sheriff. Later he worked as a night guard and detective in Price, Utah. He died a natural death on December 21, 1938, at the age of seventy-four.[2]

Matt Warner in later years

Matt Warner worked with Butch Cassidy's bandit gang known as the Wild Bunch. The gang had many hideouts, including the Robber's Roost in the San Rafael region.[3] It's possible that the Wild Bunch had a hideout near Price. It might be worth the time and effort to hunt a little closer to Price for the lost loot of the Wild Bunch, since it seems that a ride some 150 miles south to the Robber's Roost with packs of loot might have been a bit much for them. After numerous hold ups and robberies, they more than likely had a cache site close by that they utilized while they waited for the posses to stop hunting them, finally taking their loot on to Robber's Roost.

West of Price City stands a lonely structure of limestone: a pinnacle of sorts that some say was used as a guide to find where outlaws stashed their loot. A good metal detector and some time might produce a buried cache or two.

Matt Warner's headstone

THE GIANT MAN

Many treasure hunters are closemouthed and they don't often talk about what they find, except in their tiny circles. Once in a while, something slips out and then the rush is on again.

Somewhere in San Rafael Reef are the remains of a giant man laid out on a stone bed. He measures nine feet from head to foot. Two railroad workers found the giant by accident while prospecting on their day off. They told of a sword that lay alongside the giant that was too heavy for either man to lift. They reported that the sword was somewhat rusted but still retained its weight. The large man appeared to be Asiatic and his mode of dress more or less confirmed it. The area where this giant is supposed to be laid up was given to the author, but so far nothing has been found of him.

J. D. BOYD SHOOTOUT

John D. Boyd Jr., a freighter and stage driver, also known as Jack, came up from Arizona where he had been involved in a trial involving armed robbers attacking the stagecoach he drove. Once in Utah, he went to work for Brig Stewart and L. C. Maringer, who had a mail contract from Price, Utah, to Fort Duchesne, Vernal, and other points in the Basin. This contract was to begin July 1, 1894, so Boyd headed for Price and a new job.

Boyd was a bit uneasy because those he had helped convict promised to kill him on sight if they ever found him. However, his prowess with a gun, as well as with his reputation, followed him to Price, and so did some shady characters that had it in their minds to harm him.

J. D. Boyd's daughter, Ida Reed, collected his journals and other memorabilia to be housed in Special Collections at the Harold B. Lee Library at Brigham Young University. From those journals, we quote Boyd as he tells the story:

> We arrived safely in Price, Utah, on June 26th. We had just settled into our hotel, when a young girl downstairs overheard three fast-draw gang members plotting. Those men were Tom Burns, John Griffith, and Hartman. We had befriended this girl earlier, and now she heard these men say, "We'll fix that . . . Boyd! We'll shoot him on sight!" Well, this girl got word to us about this threat, and we felt like we were crowded into a corner with no way to turn. I figured they had heard about the capture of Jackson in Arizona, and wanted to sort of even things up.
>
> Picking up my shooting iron, I checked to make sure it was loaded. I then moved low out of our room, sneaked down the stairs and out into the street. When I got there, the boy who'd been bragging was there, waiting for me. My heart was pumping, as he said, "Now I got you, you scoundrel!" He went for his gun, but I was faster. A second later, when the smoke cleared, I was still standing while he was lying in his own dust.
>
> I was cleared of any wrong-doing, since I had just been defending myself. It was a close call, through, and I was glad to come out on top.
>
> Not long after this shoot-out, our other coaches and teams came from Salt Lake City, and met us there. We positioned each of these along the different lines, ready to start, and on the morning of July 1st, 1894, I drove the first white-topped stage for the Intermountain Stage

> Company out of Price toward Fort Duchesne. When we reached Nine-Mile Canyon, just above the 250,000 acre Nutter's Ranch, I stopped the stage.
>
> I then crossed the stream and climbed to the slate rock ledge. I had carried some axle grease with me for writing, and once there I wrote my name and the date on the side of the rock: J. D. Boyd July 1, 1894. I was glad I took time to do this, too, because my mark was immediately sunburned into the rock, making it become a permanent testament of my being there. The 9-Mile Canyon road was built by the all-black U.S. Cavalry soldiers. Fort Duchesne was the only connecting road from the Uintah Basin to the Railroad in Price. Thousands of immigrants used this road when it was open to the public for homesteading.[4]

The road Boyd mentions is now mostly paved, except for a few miles in the lower part. With all those thousands of wagons, coaches, and men on horseback, there have to be items left behind, hidden behind a rock or buried just a few inches below the surface. It is a great area to prospect and to metal detect. However, there are archaeological sites in abundance in the area that should be left alone and not tread on. Respect private property and please don't leave your "mark" on the rocks or landscape. Have fun exploring and respect the land and its resources.

NOTES

1. Rufus Wood Leigh, *Five Hundred Utah Place Names: Their Origin and Significance* (Deseret News Press: Salt Lake City, 1961), 11.
2. "Old West Outlaws: Willard Erastus Christianson, aka: Matt Warner, Ras Lewis, The Mormon Kid (1864–1938)," *Legends of America*, accessed April 5, 2017, www.legendsofamerica.com/we-outlawlist-c.html. Used with permission.
3. "Outlaw Gangs: The Wild Bunch (1896–1901)," *Legends of America*, accessed April 5, 2017, www.legendsofamerica.com/we-outlawgangslist5.html#The Wild Bunch (1896-1901).
4. Ida Boyd Reid collection on the Boyd family, 1839, Special Collections, Harold B. Lee Library (Provo, Utah).

Daggett County

Daggett County is "Utah's youngest county, created in 1918; it comprises an area in the state's northeast corner between the crest of the Uinta Mountains and the Utah-Wyoming boundary. It has a very small population. With the building of Flaming Gorge Dam in Red [Canyon] of Green River there is a great impetus in life in Daggett. Daggett County was named for Ellsworth Daggett, surveyor of the canal to divert irrigation water from Henry's Fork."[1]

HIDDEN GOLD AT BROWN'S HOLE

In 1825, Brown's Hole was visited by General William H. Ashley, owner of the Rocky Mountain Fur Company. After his visit, Brown's Hole "became the scene of fur company activity."[2] In later years, Fort Davy Crockett became the place to trade and sell furs for supplies or cash. The whole part of that country depended on Fort Davy Crockett. Not only did traders come to trade, but they came to drink whiskey, tell the best whoppers, and test each other's abilities at hatchet and knife throwing, shooting, and gambling. There were plenty of bottles of whiskey, knives, hatchets, fishhooks, traps, and barrels of gunpowder, as well as a few women.

In 1844, Fort Robidoux (a neighboring fort) and the cabins in the surrounding area were burned to the ground by Ute Indians.[3] Fort Davy Crockett "went out of business in the late summer of 1840."[4] Survivors told of a cache of some 1,400 $20 gold pieces being hid in the walkway of one of the main cabins that was burned to the ground. This cabin was located at Bird Springs near the South Mountain, at the west end of Brown's Hole. One source says that the coins were not part of the mountain man caches but were more likely from the Wild Bunch who used a cabin in the same vicinity.[5] It is thought that the coins are still there and just waiting for some lucky treasure hunter to get them.

CACHES FROM THE WILD BUNCH

The story of the Wild Bunch being in Daggett County is no secret. Tales of caches and hidden loot have inspired countless prospectors and treasure hunters to venture into the area, seeking those lost riches. One such story surrounds a man and his wife who were camped near an old schoolhouse. The husband would go out searching for a certain flat rock not too far from their camp, not more than a mile in any direction. He was looking for a flat white stone with certain markings on it; he had acquired a map that told of a cache of double eagle coins buried beneath the stone. He searched every day for several years, and then one day he came back to camp for his white mule and panniers (packs used on horses and mules). He told his wife that he had found the stone and would need the mule to carry out the cache because it was so large. Excited beyond belief, he took off on his horse with white mule in tow. Hours passed, then a day, and his wife got extremely worried. She went off in their old truck to find help to look for her husband. When she returned hours later, she was stunned to see the white mule standing next to their camp minus panniers. There was no horse and no husband with him. The search began, and after days of searching, the posse gave up. The old man and his horse were never found, and neither were the double eagles. Did someone bushwhack him? Who knows for sure?

THE LOST MINE IN JESSE EWING CANYON

Jesse Ewing and his lost mine are well-known in the Uinta Basin. Ewing was a tough and rough man who was as crude as any one man could be. He discovered a large outcropping of copper ore on Dyer Mountain, north of the old outlaw town of Bullionville. He began digging a tunnel to intersect with the main vein. He worked long and hard to achieve his goal but soon ran out of funds to carry on the project. It was his habit to take in partners, giving them an equal interest in his projects in exchange for cash the partners put up. Usually, Ewing would find a reason to run his partners off and this he did with knife or gun.[6] His first partner on the project was a black man who called himself "Quick Shot."[7] The second partner was a man named Coulter, with whom Ewing seemed to have had a previous acquaintanceship.[8] The third partner was slashed to death with Ewing's knife, and his body was dumped into the Green River.[9]

During the digging of the long tunnel to the main vein of copper, Ewing instead hit a very rich vein of gold. He worried that someone would come and rob his mine if he wasn't around, so he built a cabin over the entrance to his mine and tried to remove all evidence of the mine even being there. Pick Murdock, a Ute Indian, knew Ewing very well, and at one time, Ewing gave Murdock a sample of the ore from the mine. It was copper and gold and worth a great deal of money then and worth much more today.

The demise of Jesse Ewing came shortly after he got himself a woman by the name of Madame Forrestal, who was well-known throughout the mining camps. He also hired a man by the name of Duncan, who was a rowdy outlaw. Duncan and Madame Forrestal soon became lovers, and Ewing knew it. He set out to catch Duncan in the act of making love to his wife. So, one day, he faked leaving the mine for a time, but circled back to lay in wait for Duncan to show his true colors. Little did Ewing know that Duncan was waiting for him. When Ewing stuck his head around the edge of a cliff, Duncan mounted his rifle and fired, practically blowing Ewing's head clean off. The lovers than took all they could carry on Ewing's two mules and headed to a nearby ranch to report that Ewing was dead.[10] Whatever happened to Duncan and Madame Forrestal is unknown, but we do know that they had plans to come back and help themselves to the rich gold that lay hidden in Jesse Ewing Canyon. As soon as word got out that Ewing was dead, prospectors and treasure seekers flocked to the canyon in hopes of finding his rich mine. Several tunnels were discovered but not the mine that produced the rich copper and gold ore. At the time of Pick Murdock's death, a sample of the Ewing ore was sent to Denver for an assay. It proved to be very rich and demanded the price of $1,500 to $5,000 per ton.

NOTES

1. Rufus Wood Leigh, *Five Hundred Utah Place Names: Their Origin and Significance* (Deseret News Press: Salt Lake City, 1961), 16–17.
2. William L. Tennent, "Brown's Hole 1825–1871: Explorers, Traders, and Mountain Men," in *John Jarvie of Brown's Park* (Bureau of Land Management—Utah, 1981).
3. John D. Barton, *Utah History Encyclopedia*, s.v. "Fort Robidoux," accessed April 5, 2017, www.uen.org/utah_history_encyclopedia/f/FORT_ROBIDOUX.html.
4. John D. Barton, *Utah History Encyclopedia*, s.v. "Fort Davy Crockett," accessed April 5, 2017, www.uen.org/utah_history_encyclopedia/fFORT_DAVEY_CROCKETT.html.
5. Ida Cowan, interview by Steven Shaffer.
6. "WWHA Gravesite Tour: Jesse Ewing," comp. Steve Grimm, *Wild West History Association*, accessed April 6, 2017, www.wildwesthistory.org/grv-zth-ewing.html.
7. Charles Kelly, *The Outlaw Trail: A History of Butch Cassidy & His Wild Bunch* (Lincoln: University of Nebraska, 1996), 69.
8. "WWHA Gravesite Tour: Jesse Ewing," comp. Steve Grimm, *Wild West History Association*, accessed April 6, 2017, www.wildwesthistory.org/grv-zth-ewing.html.
9. Charles Kelly, *The Outlaw Trail: A History of Butch Cassidy & His Wild Bunch* (Lincoln: University of Nebraska, 1996), 71.
10. Ibid., 78–79.

Davis County

"Davis County was one of the original eight counties created in 1850 when Utah Territory was organized. Davis is the smallest county; it extends westward from the crest of the Wasatch Mountains, across the narrow, rich piedmont, the salt-marsh land, and into the Great Salt Lake to include Antelope Island. Besides horticulture and dairying, Davis County is fast growing in industries and Government installations. . . . The name Davis was for Captain Daniel C. Davis of the Mormon Battalion."[1]

THE OLD SPANISH FORT

There are a few tales of hidden treasure, gold, and Spaniards in Davis County. One such tale begins with a strange stone fort, discovered where present-day Kaysville City now stands. When settlers discovered the old fort, they thought it was strange indeed, since they assumed they were the first "civilized" folks to inhabit the region. The only other inhabitants the settlers knew of were the "digger" Indians, as they were called by the settlers, and the "diggers" did not seem to fit the mold of stonemasons. It was discovered that trappers and mountain men had visited the old fort, mentioning that it had been in ruins many years before the Mormons came to the area. The trappers and mountain men knew it hadn't been built by trappers because there wasn't much to trap in the area worth their time and effort.

In 1848, Samuel Parrish helped establish a colony[2] in the mouth of a canyon (he named after himself) between Bountiful (to the south) and Farmington Canyon (to the north). The town was called Centerville, and it was home to farmers and ranchers who grazed sheep and cattle on the foothills and raised crops on the fertile flat land below.

In 1909, a man by the name of Thomas Christensen settled in Centerville and operated a gristmill on Parrish Creek. One day, Christensen took his family on a picnic into the upper reaches of Parrish Canyon—to a flat or meadow suited for such an event. While having their lunch, Christensen noticed a cave way up the rocky hillsides that beckoned him. Being somewhat of a curious man, he headed up to the cave. Once there, he could see it was big enough for him to get inside to explore it. Once inside, his eyes adjusted to the darkness, and he spied several pieces of metal on the floor of the cave. He was looking at an old crossbow and several pieces of heavy Spanish armor. He carried out a large and beautiful breastplate and other objects to his mill at the bottom of the canyon. Here they were displayed for all to see. The people of the area decided that the old fort at Kaysville had to have been built by Spaniards. Nobody knows what happened to the Spanish artifacts, since the Christensens moved from Centerville to the Northwest. Nothing more was discovered in Parrish Canyon, but the legend persists and people still go up looking for that which is probably not there.

LOST GOLD IN FARMINGTON CANYON

A local prospector discovered a few gold nuggets up Farmington Canyon. This discovery started a mini gold rush. During this gold rush, an old mine was discovered not far from the top of the canyon. Soon afterward, another old mine was discovered in Ward Canyon nearer to Bountiful. Both mines had Spanish artifacts in them.

My grandfather, Ianthius W. Barlow, was hired by the owners of one of the mines. His job as a young lad was to drive the teams of horses that were loaded with gold ore out of the canyon and up to the railroad. This would be the first load taken out of this rich mine. Grandfather was paid $1 a day and was given his meals.

With that heavy load of gold ore and a span of horses, he worked his way down the canyon to a makeshift miner's mess hall situated up on a little bench just north of the mouth of the canyon. Leaving the load and horses on the road, he hiked uphill to the mess hall for a bite to eat before continuing on his trip. It was then that a major rainstorm came in. He thought it would soon pass over, but the storm was relentless.

He decided that he had better get the load and the horses under cover, but as soon as he hit the door, a loud rumble was heard up the canyon. It grew louder by the second and then a huge rush of water could be seen rushing down the canyon toward them. Huge rocks and trees were pushed in front of the wall of water that was estimated to be at least sixty feet high. He thought, *The horses!*

He suddenly realized what was about to happen and scrambled out of harm's way as the horses, ore, and wagon, plus other drivers' wagons and horses, were swept away and scattered all over the bottomland. The ore was lost and all the animals were killed.

As for the mine, it was never found again. All traces of it were erased from view, and the canyon was scoured clean. Now and then, someone finds a chunk of gold ore in and around Farmington Canyon. However, nowadays, it's pretty tough to get permission to hunt for gold there since most of the land is now private property.

BAIR CANYON MINING TRAIL

The two upside down crosses in the photos above are located east of Kaysville in Bair Canyon. There also is a large granite boulder with two characters inlaid with silver on its underside. The meanings of these two characters are a mystery. However, one thing is for certain, this trail leads to the top of the mountain and ties into what is now known as the "Great Western Trail." Before that trail was established, the old trail out of Bair Canyon cut around Francis Peak and went eastward into Deep Creek. This area has been established as a Spanish mining area.

Upside down cross, indicating "begin trail here"

Upside down cross, indicating
"begin trail here"

NOTES

1. Rufus Wood Leigh, *Five Hundred Utah Place Names: Their Origin and Significance* (Deseret News Press: Salt Lake City, 1961), 17.
2. "150 Years in Centerville," comp. Twila Van Leer, *Centerville City*, accessed April 6, 2017, centervilleut.net/communityinfo/history/.

Duchesne County

Duchesne County was named after a trapper named Du Chesne who lived in the Uinta Basin area around 1840. "Father De Smet may have been instrumental in naming the River in honor of Rose Du Chesne, founder in America of the Society of the Sacred Heart. [Rose Du Chesne] came from a prominent French family of St. Charles, Missouri."[1]

Duchesne County is full of historical events that range from Spanish miners and slave traders, to Indian wars, trappers, prospectors, and more. Much has been written about these subjects, but there are a few stories that haven't been written about or published. Many are told around the kitchen table on cold winter nights or around the campfire.

THE LOST BINGGELI MINE

The following story has been told and retold dozens of times, but seems to have twists and turns added to it each time. Stories should be told exactly how they are received, much like the Indians used to do in times past. They wouldn't change a single word because it would make the story null and void. Nowadays, we either diminish or embellish stories to give just enough information to entice our listeners or embellish just enough to keep them interested. Stories are best told without alteration or not told at all.

The lost Binggeli Mine has enticed prospectors for many years. The late George Thompson wrote about this mine in his book, *Faded Footprints*,[2] and elaborated extensively on it. Unfortunately, Mr. Thompson had a habit of embellishing his stories and falsely labeling photos he exhibited in his books. To get the gist of the story, it is suggested that readers familiarize themselves with Thompson's story, but that they read it with skepticism.

The late Barney Powell, a prospector for many years, searched not only for the Binggeli Mine but for the Rhoades Mine, the Mine of the Lost Souls, and many others.[3]

Mr. Powell's claim to fame is his destruction of the Indian Crossing on Rock Creek where the road passed through Rock Creek to McAfee Basin. This caused a great deal of concern among the Ute tribe as well as the people who had always accessed McAfee (a canyon that lies east of Rock Creek in the Uinta Mountains) by way of the Indian Crossing. Today, the crossing is off-limits to non-tribal members, so getting into McAfee is a bit tougher than before.

Powell wrote in detail about his search for the lost Binggeli mine:

> I'm over here at Heber City, Utah. I'm at Mr. Bingley's place.[4] We've been over two or three times to make the recording that we made the mistakes on. It didn't record, so we're telling the whole story. Bing is telling his story, and we are giving all the information that we have concerning this mine.
>
> This is November 3, 1981, on a Tuesday. It's quite a story. It's true, every bit of it is true. I met Bing up there as I stated before, and we looked for it [the lost Binggeli mine] for several years. One of the fellows that packed the acid with his horses for years couldn't find the camp where they were camped. It was down just east of the Iron Mine Lake. The Iron Mine Lake sits a couple of hundred feet up on the side of the hill, and there's a creek that runs down through the bottom where the water drains from all directions into that creek and they had their camp, I'd say, somewhere around 800 to 1,000 feet east of Iron Mine Lake. It was east of Iron Mine Lake that Bing, as he stated, was buggin' timber; and as he came from the buggin' of the timber, he fell through a hole. He couldn't see it, and there were rotten logs over the hole. They had been there for 100 or 200 years, we don't know. The log gave way with him and when it did he grabbed a little quakie, nice, right there by the side of the hole. We have stated this hole was about 4 feet wide, about 6 feet long, and it had quartz on both sides of it. In the middle of it was quartz, gold, sugar gold, and different kinds of gold in it. There were steps on an incline, it wasn't straight down. Steps were hewn out of the white quartz and it went down into the hole. Bing say it might be 100 feet deep, but he doesn't know for sure. It was snowing and he couldn't tell just exactly how far down the thing was. So all we're doing is just guessing because he threw rocks and logs down into it and couldn't tell how deep it was. I have always visioned this to be maybe as far as the river, 2,000 feet thickness of this vein. I know we are close to it. I know we will find it,

and we will find the whole thing. This very rich gold mine which goes back hundreds of years, maybe even to the time of the creation.

This mine was found by the Indians hundreds of years ago. They knew they found something, but did not know what it was. They called it money rock. This story goes back to 1846. Thomas [Rhoades][5] came from Illinois or back that way. He had heard something about California and there with the Mormon Battalion. He and the other Mormon Battalion were the first to discover gold there [at Sutter's Mill].

I think Thomas Rhoades was a member of the Church, and through persecution was driven out of Illinois. I don't know where he married his three wives, but he did have three wives. When he came through Salt Lake Valley, he went on to California. He did not know that that was the place the Lord wanted Mormons to settle in, so they went on to California.[6]

The Rhoades and the Mormon Battalion were the first to discover gold at Sutter's Mill.[7] Brigham Young found out about it because some of the Mormon Battalion stayed at Sutter's Mill and worked there. Thomas Rhoades stayed there for four years. Then in 1847, Brigham Young and the Mormons migrated from Nauvoo to Salt Lake City. When Brigham Young came into the Valley he said, "Yes, this is the place."[8]

Four years later Brigham Young started to build the Temple. It was about half completed when he needed gold to put in the Celestial Room and throughout the Temple, and to paint the Angel Moroni which is gold plated. These are some of the things that we know.[9]

Brigham Young said, "It is better to feed the Indians than to fight them."[10] He sent them flour. I can remember in Price when the Indians came through the Uintah Basin to Nine Mile, they would come up on those flats out from our old farm. They would spend two or three weeks shooting prairie dogs. The Squaws would come into town and bum flour and anything else to eat, and we would give it to them. I was just a small boy at the time. Brigham Young said, let's feed them instead of fighting them, and that broke their hearts. Chief Walker was the chief of all the tribes in the Uintah Basin.[11] So when this happened, he decided that this gold or what he called 'money rock' could be used for the Temple.[12] He went to talk to Brigham Young. After they made their treaties the Chief said, "You send the man and I will show him where to get it." That is why Brigham Young called Thomas Rhoades, he was a rugged outdoorsman and he was able to go and do things like that.

Brigham Young and Thomas Rhoades along with Chief Walker made a covenant never to tell anybody where the gold was. The reason the Indians did not want to do that is because from about 1600 on down 100 years or more the Indians were enslaved by the Spaniards.[13] The Spaniards came up the Green River and found the Indians around Vernal or west of there. When the Indians saw them coming, they thought they were the great white gods come to rule.[14] They knelt down and worshipped them, they even sacrificed some of the Chief's children to the gods to try and tell them who they were. Many others were sacrificed by the Spaniards, which they claimed was to the great white god. This was learned in South America where the Lamanites and the Aztecs and Mayas performed that ritual. I think that is what started the whole thing. So, the Spaniards enslaved the Indians for a little over 100 years. It is for that reason when the Indians freed themselves, they killed the Spaniards. They threw them in holes and trenches. The Indians never payed *[sic]* attention to the gold, but the Spaniards made them dig the trenches to get the gold. When the trenches got deeper, they called them holes, which we call a shaft. I, B. G. Powell, would like to go back and tell you how this started.

About July or August Uncle Rob appeared to me. I was down in my basement taking a shower and all at once he appeared. He spoke to me and said, "You have been looking in the wrong place." That is all he said, and he then just vanished. I saw Uncle Rob just as plain as the sun comes in the morning. He and I had spent fifty years looking for this mine. He died in about '57. We would go out there in my pickup and prospect for gold. We never did find it but Uncle Rob lived out there in the reservation and he saw the packs of gold, he saw much of it. He knew Rhoades was bringing it in. He saw him many times with pack saddles of it, and he knew where he had come from—from up in the Uintahs. But Uncle Rob could never find it.

When Thomas Rhoades came from California when Brigham Young asked him to four years after Brigham Young had landed in Salt Lake City, he brought four bushels of gold to give to the Church; and the Church used that in the decoration of the Temple. Brigham Young had a mint where they minted the money. Thomas Rhoades was chosen by Brigham Young and Chief Walker to go out and get gold. They had made a covenant never to tell anybody. The Indians thought maybe the white man would enslave them again or kill them.[15]

I have talked to two or three Indians and they have said if Indians get gold, white men will kill them. It is for that reason that Indians would not sell gold. They do not want anything to do with it. That is one of the things that have happened, and I know that it is true. Thomas Rhoades was chosen by Brigham Young and ordained, and Chief Walker told him to come to a certain place where he had a Buck meet him out there to show him where the gold was. He told this Buck Indian that if anything happened to Thomas Rhoades, the Buck would be killed. The Buck protected him, and he went several times. I think at that time they were all living in Kamas. My father was born there, and I am sure that Thomas Rhoades was living there too.[16]

While in Kamas, I would like to relate this story. My grandfather and Kail Rhoades were hunting deer up in the Uintahs by Wolf Creek or Soapstone, and they killed a big buck one afternoon late. They couldn't take it to camp, so they hung it in a tree and cleaned it. The next morning they went back and got it. Several years later they moved to Price. In Price they had a party for all the people of Price, the settlers, and Kail Rhodes was at that party. I am getting ahead of my story; but anyway they asked Kail to tell his story and he did. He said to my grandad, "John, do you remember when we killed the big buck up at Wolf Creek or around there? From there it is not far from where I get my gold." That is what he told him at this party. Thomas Rhoades went and got the gold for years. One time he got sick, and Brigham Young needed the gold. Brigham Young and Chief Walker chose Kail Rhoades, the son of Thomas, who was about 18 or 20 years old when they chose him to get the gold. When his father got better, they both went and got it up until the time that Thomas Rhoades died; and then Kail continued to get it. Later they moved from Kamas to Payson. A lot of our people, my grandad and others are buried at the Payson cemetery.

Uncle Abe Powell was a trapper and hunter. He went up Spanish Fork Canyon trapping beaver. He went clear to the head of the canyon; and when he got to the head of the canyon, he saw beaver tracks going over the hill down into somewhere, he didn't know the name of it at the time, but today it is called Soldier Summit. They went down into a place called White River, and found lots and lots of beaver down there. So he became curious, he wanted to know where that river went. He went down to Colton, there it goes into the Price River. The Price River heads at Schofield Clear Creek and even farther up, so he followed that river on down into the valley. When he went to what is today Castle Gate, he saw great veins of coal sticking out of the mountain. He reported that and went on down through Helper. That wasn't too much of a good settlement for farming, so he went on down past the Blue Cut and what they called at that time Castle Valley. Down in there below the City of Price, the south end of it, he built a cabin and stayed that winter. My grandad settled over half the valley from the main street in Price on south clear to the river. You see that started the whole thing.

Well, Kail Rhoades came along with them. Kail Rhoades settled just below the Blue Cut, that is half way between Price and Helper. He had about 160 acres in there and he built a cabin. One time, I want to tell while I am thinking about it, Kail Rhodes went to Price to get some groceries and met my grandfather, John Powell. He said, "John, come on up, I've got something to show you." So my grandfather went up in a day or so in his white top buggy. Kail took him out to a cellar, and there were several pack saddles full of gold. My grandfather put his hand in the gold nuggets and sifted through. That is one story that we have to believe.

My father was born in Kamas and they must have lived there for quite awhile; and that made it good because many of the settlers in Kamas at that time testified that Thomas Rhoades would go up Beaver Creek, which is east of Kamas, and go on up somewhere they didn't know—they tried to follow him but they lost his trail. He would go up into right around what is known as the Iron Mine or Soapstone, and there I think he got gold.[17] He would also go over to Wolf Creek and down Wolf Creek about 40 miles. East of there is the vein of gold. They found gold in several places along that mountain. Clark Powell just last summer between Rhodes Canyon and the Duchesne River right up on Lightning Peak discovered gold, he showed five nuggets to some people or kinds that were prospecting. He told these kids he had found Kail Rhoades's gold mine.

I would like to give a little history now. Brigham Young died in 1877. Thomas Rhoades died about 1860 or somewhere in there. Caleb Rhoades died in 1906. The reservation was thrown open for settlement in 1906. Just before this, Caleb Rhoades died. He told my grandfather, John A. Tuttle, as soon as the reservation is thrown open, we will locate it. We will be rich. We'll never know what to do with the money it is so rich. So that is the story that Kail told. Right after that Kail got sick and died suddenly. Kail Rhoades at this party in Price swore up and down on the Bible "I kept my oath and to this day I have never told any may where it is."

Kail Rhoades married Aunt Martha Powell, my grandad's sister. They lived for a long time around Payson and Kamas. I don't know too much of their history there, but I do know this, that they had a boarding house up at Tinnick.[18] Tinnick was going big and she run the boarding house. Kail would go out into the mountains to get gold. Blackhawk[19] told Uncle Rob at times that Kail would come out there and stay with him as high as a month or two months. When he got ready to go get the gold, he would wait for the full moon to go. Kail would never travel out there in the daytime. He would travel always at night, in the full moon when the moon was good and clear and he could see where he was going. [Mountain Sheep] says that he stayed there with him for a long time. The Indians called Kail Rhoades a "*heap wino friend*," which is brother. They would do anything for him and he would do anything for them. He loved the Indians and they loved him. [Mountain Sheep] told my Uncle Rob that Kail would leave his camp at sundown, just as the moon was coming up, and would travel in a northern direction. Now [Mountain Sheep] had his camp at Pigeon Water. Pigeon Water is north of Mountain Home and it is east of Rock Creek. You go up a hill there 3 or 4 miles, then when the road turns toward Mountain Home, down a little trail there is a sign that says Pigeon Water 1 mile north. There is a spring up there and that is where [Mountain Sheep] had his camp. When he would go for his gold, he would go in a northern direction.

He could have gone up into Dry Canyon, up into the head of it there at the formation to carry his gold, or he could of went farther on over into the big ridge and above Peterson Ranch and all in there. The Indian Ranger, Curry, told us that he saw trail blazes on trees up above what we call Cottonwood Grove on Rock Creek, it is about one mile and half up the river from where Farm Creek crosses the Duchesne.[20] Then you go up that canyon just west of there, and you could trail him by the blazes for a mile or farther. Kail said one time he got all the gold he wanted out of a pine tree. There was a pine tree that fell over leaving the roots exposed and he got all the gold, more than what he could take, out of the roots of that tree. That is one of the stories.

While I am in this area, Earl Wright's father and Leland Wright told me two years ago on decoration day at Tooele the story of how his grandfather had run cattle up in there. All this country clear up to Moon Lake and above there, way high into the Uintahs over to Mirror Lake and all through there was Indian territory at one time. The government took it away from them and moved them down into the lower part, they are down in the foothills. Anyway, he had to make a deal with the Indians to run his cattle. One fall he was running the cattle out and he was under the ledges, he would throw rocks down to run the cattle out. He picked up rocks that were not too heavy, but he came to a place where he picked up heavy rocks, they were really heavy. He took 2 or 3 rocks and put them in his pocket. He was amazed at how heavy the rocks were. The next spring he went over to the banker in Kamas, showed him the rocks, and the banker put acid or something on them and said they were pure gold nuggets. He wanted to know where he had gotten them. So he told him where he had found them, but he didn't go back there for a few years. I have talked to 2 or 3 old sons of the old timers and they told me that if they would have taken the gold that was in there, the Indians would have killed them because they didn't want nobody touching it, so that is the reason that Mr. Wright didn't go back and get the gold because he was afraid. They would have taken his right to graze cattle away from him. Earl Wright's father says I can see Peterson's Ranch down on Rock Creek, they have fish ponds there. You can see it just as plain as plain, I've been there several times and I can see where I picked up the gold under the ledge. We have that all staked and it is all filed with the county and BLM.

We are hoping next summer to go in there with a cat. The ledges have deteriorated and fallen down over the bottom of them. We have got to have a cat[21] in there to dig them out. That is the only way we will find this gold. So that is the reason we cannot find it. Anyway, over the big ridge into the east end of Farm Creek I talked to Ron Christensen, the cattle man. He was telling me that when the built the Indian fence, it is right below the bottom of the big ridge that goes through there. That is the part they gave to the Indians, it is down lower. They built a fence clear along from the north fork of the Duchesne clear over into Rock Creek, and then up Rock Creek a ways, and then over into Moon Lake. They built the fence with the help of the Indians. The Indians knew where the Spaniards had their smelter. They had a mill there where they refined the gold, they crushed it, and washed it out and took the mud out. Down about a mile I have seen hundreds and hundreds of places where they have ponded the water, and they made ponds just like Kennecott. Great big round ponds, one below another, and there is a big sagebrush growing that is 5 or 6 inches at the stalk. Anyway, just above the smelter is about 2,000 feet inside of the forest. Those Indians working on the fence knew where the fence was going. After they killed the Spaniards, they must have had a lot of gold bars left there at the smelter. It is just above what they call Winchester Flats, and that is in Indian grounds, but a little farther up you hit the forestline. The Indians buried the gold bars that were left by the furnace when they killed the Spaniards. I think that is there yet today, if we could find it. I think the Indians went up late at night and took the gold bars down and buried them in Indian grounds. They knew they could not get them after the forest fence had been put in there. So today, you can see where they had a smelter, It's under a big rock slide, they had rocks for the furnace, and they smelted the gold and milled the gold right there. I honestly believe that they did.

I wanted to come to the time when Kail Rhodes had the boarding house up at Tinnick, Utah. He left there to go out and get gold and had been gone for a month or so; he came back unexpectedly and he found a man in the bed with his wife, Martha Powell. He threatened to divorce her, but she pleaded with him to forget about it, so he did. He went back out and only stayed a day or so and came back and found this man in bed with Martha again. He divorced her and kicked her out. As I understand it, she left there and went to Green River and then to Grand Junction, and this guy that was with her followed her. He later married her somewhere down there, and move to Grand Junction. I know that she has several sons and maybe daughters. These sons work as engineers on the Denver & Rio Grande railroad at Grand Junction, Colorado. That is the history of why I wondered for years, I read the book and found out why they did it.

Kail sometimes before the nineteenth century[22] sent some of this gold or went to Washington D. C. and he offered to pay the national debt if they would give him the rest of the gold. The debt at that time was about $50 million, and they turned him down. The rotten crooks in Washington formed what they called the Florence Raving Mining Company. They came out and tried to find all the laterite, the [gilsonite], and everything else; but they couldn't find the gold, so one day they offered Kail $10,000 to show them where the gold was. He said, "Why I get more than that every time I go there, why do I want to sell it to you?" He couldn't break his covenant; he couldn't have sold it to them no matter what he wanted to do, because he had made that covenant and he had to live up to that covenant. So you see those are things that go to prove the worth of many things that took place.

Another story I want to relate to you is—Uncle Rob, when he lived in Cedar View, had a store and he sold stuff to the Indians, groceries and stuff like that. The Indians would come and buy all the vanilla extract he had. He would try to buy as much as he could, but they would buy every bit of it and get drunk on it. So while they were doing this, he kindly put the pressure on them to get them to tell—it was only the chiefs that were able to buy the vanilla extract. I found this out a couple or three months ago from Cora Todd, his daughter. She told me this was one thing I don't think Uncle Rob ever told; I'm going to tell you. And she told me the story about the vanilla extract.

I'd like to go back to Kamas and tell you about the hardship that they had to put up with up there. The snow got so deep. In the winter they had to take their grain down to the gristmill at Heber to have it ground into flour. When they would leave Kamas, the snow, I guess, would be three or four feet deep, maybe deeper. They couldn't pull the wagon in the snow, so they would take the horses to go along and make several trips back and forth to break the snow. Then they would come back to the wagon and pull it down as far as they had broken the snow; and that is how they got out of Kamas to go to the grist mill to have their flour ground. So if that ain't quite a deal, I don't know what is.

One time Uncle Rob and Bishop McMullin and two other men decided to go out and look for the gold. Bishop McMullin was the Bishop at either Price or Wellington, I don't know which. They came out to Duchesne. Mountain [Sheep] was one of the chiefs, and he had a ranch down, I would say, around Bridgeland, it was down the Duchesne River.[23] Anyway, they went down to this place because Uncle Rob knew him. They talked to him and said they would like to have him show them the gold. He said no. He says, "all right" but he crossed his fingers, "We will give you half of it and make you rich." So they finally talked Mountain Chief into telling them. "My son is at Heber City and as soon as he comes home on Saturday I'll send him, and he will show you where the gold is." So they stayed there a little while, and got talking among themselves and said, "Why, if it is that easy, let's go get it ourselves and not wait until he comes home. Let's get it all." So they went looking and tried to find it, and couldn't find it. When they came back to Mountain Chief's far, he would not have anything more to do with them. I guess Mountain Chief had a Buck follow them to see what they were doing. So they never did find it. Uncle Rob told me that was the only time in his life he had a chance to find it, but they blew it by going and trying to find it without Mountain Chief's permission. And that was another of the stories.

Another one of the stories was—when Kail would go out there, they would follow him, and follow him for months. Valinger and Witmore were the bankers in Price, and they followed him from Price. They were sure they were going to find it and really get rich. Kail would just run them around, run them crazy, and never go near the mine. One night Valinger and them camped in kind of a valley, and there was a draw down in the bottom that ran up a side hill across the draw. They were practicing with their six shooters shooting over at the other side of the draw. There was a high sagebrush about 6 or 9 feet high. So Kail [Rhoades] knew where they were and snuck up on them; when they laid the six shooters down at their camp, they walked down across the gully to the other side, and Kail went up and swiped their six shooters. They couldn't figure out what happened to the guns, they were just amazed. So finally they gave up looking for Kail and went back to Price. After they quit following him, Kail went to the mine and got all the gold that he wanted. When he had gotten the gold, he went home to Price. A day or so later he went to Price to the bank and took the guns into Valinger and Whitmore and asked them if they knew who they belonged to. They said, "Yes, where did you get them?" Then he told them the story about looking at where they had been shooting and he went up and swiped their guns. They couldn't believe it. So that is just a story of how they tried to find the gold and trap Kail [Rhoades].

I forgot to tell you about after Kail divorced Martha Powell, his wife. He sponsored an English girl to come over here. She was so grateful to Kail [Rhoades] that she flirted with him, and finally Kail married her and they lived for some time there below the Blue Cut. Finally before Kail died he tried to make a map, and he did make a map. She took it and went out into the Uintah Basin for 2-1/2 years trying to find the gold, and Uncle Rob saw her several times looking for the gold. She would not cooperate with him or anybody else, and she never did find the gold.

Now I want to tell you about 1932. I had been up in the Uintahs a lot of times before; but this particular time I met a man up there and he was prospecting. He said to me, "What are you doing up here?" I said, "Well, I am prospecting, I am looking for the Kail [Rhoades] gold mine." He said, "I found it, I'm not looking for it, I found it." He told me the story about when he was bugging timber for the forest service in about 1932. One night late, in the fall, it must have been November or December, they came from where they were bugging the timber. They had a camp down about 1,000 feet straight east of Iron Mine Lake right down on the creek, and they started to leave. He didn't know where it was, he said they left and he stepped off a little ledge and fell in a hole. He said he grabbed a little quaker at the side of it, and it saved him from going down in the hole. So they uncovered the hole, and what did they find but steps going down into this hole on an incline, and it was about 4 feet wide and about 4 to 6 feet long, and there were old logs about 30 inches in diameter and they were covered over this hole. He stepped on one of these rotten logs which had been there for years. The Indians had put them there to cover the hole. It had started to snow sometime before they got down to this hole. He said it was snowing so hard by then that you couldn't see 10 to 20 feet in front of you. His partner went on by it and didn't see him fall through the hole. He missed Bing, and Bing was just then crawling out of the hole. After they realized what they had found, his partner said, "Bing, there is a curse on this, we will die if you do anything about it, just forget about it and leave it alone." Bing was bull headed and said no and went ahead and uncovered it. He found a white quartz vein on the side of it, on the ends of it, about six feet or so, was pure gold in quartz, and Bing tells that he threw rocks and sticks down in it and it looked like it was about 50 to 70 feet deep. He said it took a long time for the rocks to hit bottom. It was so dark and snowing that they couldn't tell, and before they left this place there was a foot of snow on the ground. They cut trees, covered it over, and covered it with leaves, and then went on to try to find their camp; and they were quite awhile trying to find their camp. When they found their camp, they went up and down the creek to find it, there were about 45 men working there and they had the tents down and were starting to move. Bing says he cut several pieces of gold and filled his jumper pocket. It was so heavy he could hardly pack it. They were two nights and two days getting out of there, I think he said it took them five days to get down to the Stuart's ranch. They were about five days getting down there. They were all wet and had an awful time getting down there. I know that they found it.

Well, I thought Bing had died. I read in July that Bing [Binggeli][24] had died and I didn't look to see his age. But come to find out, I had been trying to find Bing Bingley for a long time, it was his son had a heart attack and died. I went over to Heber to file some of the claims and while in Heber I asked the recorder if he knew Bing Binggeli. "Yeh," he said, "He lives right up Sutter Creek." I got on the phone and called him. He was like kid with a new toy, he was tickled to death to meet me. Well, anyway I have looked and looked and never could find it. After this vision that Uncle Rob had showed me, I went down into this area, it is off of the Iron Mine Lake Road about 4 miles on a logging road, which is one of the roughest roads I have ever been on in my life. I went down pretty near to the end of the road, I never got out of the pickup, but I felt the most deep depressed feeling. I was just scared to death, I just felt like there was Indians or somebody who wanted to take my life. I turned around in my pickup, I didn't get out, I got out of there, and I had an awful time getting out. I wonder for a while if I was going to make it. This happened somewhere about the fifteenth of June. So somewhere about the fifth of July I got Dave Horn and six other men and Dave's son; I told them to come on, they can't kill all of us so we're going to go up and try. Where I had seen this vision of this mine I went up and parked, pitched our tents, walked over 1,500 feet west into a little swale. I said this looks like the place, so we went down into the swale. I had a metal detector, Dave Horn had a metal detector. I am totally disabled, I have a hard time getting around, but I can crawl over logs and places. We got down about 400 feet from where we staked the claim, and we hit the reading of gold.

This metal detector has a green light and a red light. The green light detects gold, and the red light detects iron. When you lift it about two feet off the ground, it quits. We found this reading around an area of about 100 feet. This is where the Spaniards brought the gold up. One of the guys sat on a pile of rocks just east of this place, and he says, "Hey, there is cold air coming out of this hole." So everybody ran for it. We took some dusty dirt and threw it over the place it was coming out, and for about 4 or 5 feet air was coming out blowing the dust we had thrown on it up in the air. Well, that's all it took, we started digging and we found that there was a shaft down into tunnel. When the Spaniards spoke of holes, they were talking about what we call shafts. It was about 20 feet down, and around that shaft was pure quartz, and we couldn't get into it for the rocks that had been thrown in. Some of them were 400 or 500 pounds. We lifted a lot of them out. But in the meantime, I went over and got more interested in this place where Earl Wright's father had found the gold. I wanted to go over there and stake a bunch of claims. We staked at this time about 150 claims in this area. I left there the night of the 7th to go to Salt Lake to attend a court case on the 10th at 10:00 a. m. When I left, I left my pickup and the tent all locked up.

I don't know who it was, but somebody went up there on the 8th. This man, Taylor, went up to look at it and he met a pickup about 1/4 mile coming out of this road, which is a dead-end road. The sucker had stole all my stuff, about $2,600 worth, the metal detector I had, my 22 Remington Automatic, and about 10 or 12 other items. They didn't break in, they just took what they needed. I had two barrels of gas, about 110 gallons, and they took what they wanted of the gas and threw the rest on the ground. And that is the history of that.

This Taylor went out there on about the 8th with his boy. He was more interested in trying to dig the hole out and get down in there. I know there is gold down in there. We have finally dug the hole out and it looks good. We can see a shaft about 6–7 feet high and 4 1/2 feet wide, and that is the hole that [Mountain Sheep] spoke of.

We started to stake out claims, and we staked about 75 claims in there right then all in the area where we thought the gold was. We definitely knew there was a tunnel because the air was coming out somewhere on either north or south because there is another opening where the air is going in and it is coming out at this place. Now this Taylor got snooping around, and he found a tunnel way down at the ledges that was open and maybe that is where the air is coming from.

When I talked to Bing Binggeli over to Heber, he told me, "Barney, that is right in the direction of where we came. I am sure that somewhere along the line of that tunnel you will find that rich vein of gold." We have tried to get down in this tunnel, we have pictures of it, it shows it. We stuck a bar down in the hole that Taylor and Todd had drained out, about 6 or 8 feet, and we couldn't reach the bottom.

Todd and Taylor figured old Barney Powell don't know anything about this, so we'll shoot this hole out. They couldn't get into it, it was full of white quartz, about 3 feet. As we dug the shaft down, we found wood pulp on the four corners about 4 feet square, and the dirt still looked like it had boards on all four sides of it. What the Indians did when they killed the Spaniards was throw their bodies in the holes and trenches and buried them. We didn't find any Spaniards here, but I think when we find the last hole, we'll find a lot of Spaniards. We'll find their armor, their bones, their skeletons, and everything. What Todd and Taylor figured to do since old Barney don't know anything about this is go and uncover it. They took some powder but couldn't get it down because the quartz had only about 1 1/2 to 2 feet hole, and they needed a bigger hole to get down in it. So they took the powder and shot this hole and fractured the whole thing for about 30 to 40 feet each way from where they had done it. We got a backhoe up there from Mr. Wyler from down at Marrian,[25] which is just north of Kamas, and he dug out part of the hole. He broke the backhoe down in October and had to take it to Kamas. He had to wait a couple of weeks to get parts for it, and also a valve and a piston broke in the motor. He said, "Well, I have to have a cat up there to move the big rock." He couldn't reach down in the hole to get some of it out. He cleaned it out, and we have pictures that shows a tunnel. We took quartz out of there that weighs, I'd say at least 6 or 7 tons, white quartz about 3 feet thick, and that was what was laying over the tunnel. You can see it on bout ends, the quartz is still in place but fractured. It has to be dug out with a backhoe. It was getting so late and it snowed, so I stayed up there and we tried to get him to come back until it snowed about a foot and he couldn't get in there. So he says, we can't get a cat in there. We couldn't get anybody to do it, we ain't got the money to pay for it. We offered to make good concessions to them if somebody would come and do it. Wyler came and did it for free, and he will come back.

"This tunnel we have uncovered is definitely the old Spanish workings. The Spaniards dug the tunnel and they took the gold out. That gold in that tunnel must have been at least 6 to 7 feet high and it was about 4 to 4 feet wide."

Now I'm going to ask you a question, what did the Spaniards use for lights? In about 1600. You have seen shows where they had oil torches.

This tunnel is blacker than cats. We have taken out pieces of soot off of the rocks that is 1/4 inch thick. They used torches for lights, that is all they had, and that is what makes it so black. I think the Spaniards did was every 500 or 600 feet or so they opened a place to the top where they brought the gold out. After Todd and them cleaned it out, they couldn't get down into it, but on the side was four drill marks. They talked to some engineers, and they told them that was the way the Spaniards did their work. They drilled from the bottom up and made their hole to the outside, and then they brought their gold out of that hole. That is why we got a metal detection just out from the hole, because they piled the gold out there and some of it had quartz on it. They knocked the gold off and threw the quartz away, and there is pieces of quartz in there. It has been maybe 200 years since they did this, and that goes to prove another thing. That is the tunnel. When you get out of this area, about 100 feet square, the gold does not register. But it does register the green light on this area around about 150 feet, and it is very strong and it shows that was what they did.

I think we've got to go at least 600, 800, or maybe even 1,000 feet in a northern direction where there is another depressed place, a sunken part of that tunnel, and then it shows that there is a ridge going down about 100 feet long on a slant and on up to the top of that ridge. This ridge is about 50 feet high, and at the end of that is a 3 to 4 foot pure white quartz vein running right across it, it's about 40 feet long and 4 feet wide. I think they didn't get to that quartz. I think down 30 feet from that depressed hole is the place where we are going to find pure gold and quartz in that area. I wouldn't be a bit surprised if we find Spaniards, bodies, helmet, and all that stuff right in that depressed hole. I think that is where the Indians put them after they killed them and put them in a hole, as the Indians called it. They are not holes, they are shafts. Those are things we have found, and I am sure as anything that we find the other hole and we'll find it by drilling drill holes every 10 to 12 feet. If we stay in the line of that, we will find where we are going, we'll find the place where Bing Binggeli fell through the hole where it was pure gold. That gold was not 6 or 7 feet, it was 60, or 70 feet deep, it could go down a lot more than that. That is very very rich.

These steps are hewn out of the quartz so they could go up and down out of the hole. They didn't have any other way of doing it and that was the easiest way for them to do it. They had chisels or something, I don't know what the way that they dug the holes. Bing didn't dare go down in the hole. After Bing got out, they didn't have the money to go back up and locate it. They waited two or three years and when they went back, they couldn't find it. I was up there fifty years ago, I was up there 25 years ago when I met Bing Bingley. You know, that country has changed, so many trees, aspen and cottonwood, have grown up in there until I tell you it fools ya. I can't find yet some of the places I staked 200 claims in there 25 years ago, I can't find a thing of them. And that is what puzzles me, and that is what puzzled Bingley.

Bing looked and looked, he took ropes and went up there to try to find it. Bill Nunley was the fellow that was with him. They have been there dozens of times. I've been up there six or seven times and met Binggeli up there. You just have to go on those kind of things. He would not have come up there again if he had not found it. I know he did, and I know we are right close to it. I just hope there might be two veins running through there. We'll just have to wait until spring to get in there, and then we will have all summer to get in there and develop it to see what we got, and I'm sure we are going to find it.

Two of these engineers told me, "Barney, you've got enough gold in there to pay the national debt." They just announced over the radio, television, and the paper that the national debt is over one trillion dollars. And these engineers say to me I have enough to pay the national debt of over one trillion dollars and still maybe have one or two trillion left. It's rich, it's solid, and it runs for miles through that country. We have got a stake for 7 or 8 miles, and a lot of these claims we don't need but until we find where it goes and follow it, we are going to keep those claims. Nobody can touch it.

The ranger told me yesterday, the 21st of October, 1981, that when we find the gold he will let us build a road down into Iron Mine Creek which will take us right down to the Duchesne River. That way we can get up there any time during the winter. We can keep the snow off. We can get in there and develop it, and we can work it year round. But the other way to go from there over this rough road for 4 miles and then down the old Kamas road that takes us out of there, we cannot—it is rough, it is awful, just hard as everything on the truck, and the other way we can do it. He told me he would close the area just as soon as we get the gold. And it ain't going to be only a day or two in the spring when we find the gold.

I'm sure even where the Spaniards have cleaned it out, they only took the cream, there is lots and lots of gold on roofs, rips, and on the floor, they spilled some they couldn't help but spill it. That is what we are looking for. That's what I think Todd and Taylor wanted that hole for. They wanted to get in to the ribs and they would have cleaned up a million, two, or three million dollars just on gold on the ribs and on the roof. That is why they shot the hole, to try to get in there, but it backfired on them and they couldn't get it open; in fact, they caved it in. That is why we are having so much trouble now getting it open. We can't reach it with a back hoe because it got into the deeper hillside and we have to have a cat in there to move some of the rock out so the backhoe can get in there to dig it out. So that is our problem, and it's getting so late, it is now the 21st of October. I just moved my trailer out yesterday, the 20th, and there is about 6 or 8 inches of snow on this old road up here and on the other road coming out. It is about 10 miles down to the Y.M.C.A. Camp, which takes you to the Mirror Lake road, and on the sides of the road there is about 18–20 inches of snow. There was 8–10 inches of snow on the trailer, and there was pert near an inch of ice, so you see there has been a lot of snow fall. It has been predicted that there will be a lot of snow fall again today and tomorrow. We wouldn't be able to do anything we want to do. If we could get a person to drive a cat or back hoe, I would pay them two or three times their hourly wage to get them up there to do it. When we get this, we will have enough money to buy 2, or 3, D8's, we'll buy graders, rock crushers, drills, anything that we need. We'll have lots of money to buy that stuff with, and we will go in there and build our own road ourself—and we will do a job that we will really be proud of, and it will help us to pay our tax assessment on our claims and we will have it really good.

I have left my trailer out there pretty near all summer until I brought it out to the Iron Mine here a couple of months ago; he let me take a horse and go around to the different places, and I left the trailer in his yard there. He told me this story, his grandfather rode up to Wolf Creek with Kail [Rhoades] some years ago. When they got to the top of Wolf Creek, they turned around, Kail [Rhoades] looked down the canyon and told Dale's grandfather that he had just rode over enough gold today to pave all the roads of Salt Lake City. It's there, they haven't taken it out, it is still in the ground there yet. He knew where it was, Kail [Rhoades] did, because the Indians had showed him the place. All these old Indians that live in Fort Duchesne and around know where it is, but do you think they will go get it? No sir. Do you think they will tell anyone else? No sir. They have got a covenant just as much as Kail [Rhoades] and Thomas [Rhoades] made with Chief Walker and Brigham Young that they won't tell. It is like this one told me this spring, that "If Indian get gold, white men kill Indian." That is just about what they will do. So that is the reason they don't go get it. They are afraid of it. They say that they will have it for their happy hunting ground, and that is what they want to keep it for, so they will have it in the hereafter.

Come to find out Brigham Young or somebody baptized Chief Walker.[26] Chief Walker is a member of our Church, [Mountain Sheep] is a member of our Church, and that is another reason why they did it. After Brigham Young died, the Indians cut them off. I think Chief Walker was dead at that time. I think if he had still been alive, he would still have let Kaila [Rhoades] take the gold to the Church but somebody else got in charge, and I don't remember the chief's name but he took over, I think it was his brother or some of his kin took over as the chief of the tribes.

I know, I am bearing my testimony that I know that the Lord has held this for all the time he's waited, and now I think is the time the Church needs this money. I have covenanted with all my life, with all I've got that if the Lord will reveal it, and I think he has revealed it. I think he has opened the way now for me to be one in His service, to be an instrument in His hands in developing this mine. Why would Uncle Rob have appeared to me, and he has been on the other side, and the Lord has chosen him to come back and talk to me and tell me to do it. I am not a visionary man, but I believe in those things and I have had other experiences. I have been to the other side one time before. When that guy shot me at the mine, I was dead for 16 minutes. I went beyond the veil, and I know what it is, I know where the veil is. I think the Lord has opened this up, that he has chosen me. I have been in mining all my life, I know how to mine, I know how to do all these things. I had a coal mine that was worth a million dollars one time. Alvy Vanwagner shot me, and old Ritter through the receivership and First Security Bank took that mine away and sold it for less than 1/10th of a cent on the dollar; and that is the way they treated me and that is the way they done. I know that I have been faithful in the Church, I have sent all of our family, my wife and AI and my four sons, on missions, and we are 100% in the Church, so I pray for these things. I have made a covenant with my Father in Heaven that if He gives me this mine that I will give the Church all that they need, every bit of it if they want it.

That is the covenant that I have made, and I pray to Him night and day to fulfill this great privilege, and I think that He has chosen me to do it. I hope that He has, and I hope that He will give me strength in the spring to go in there and open this mine and to develop it, and to get out the things that are there. I pray for it every night and every day. I am sure that when He sent Uncle Rob, that was His purpose, to get me to go and develop it and bring it forth in this day and age of our life.

Now I am going to get Bing Binggeli to tell his story. He knows several people that have seen the gold. Bing Binggeli on his dying day,

> if he does, will testify he hasn't changed it a bit. He has found the mine and he will tell you how he found it, that it was snowing so heavy that they had over a foot of snow when he fell through the hole and that by the time they got out of there they had from 2 1/2 to 3 feet of snow. As I stated before, in about 1935 or 1936 I was up there looking for the gold mine and I met Bing Binggeli. He asked me why I was there, and I told him that I was looking for the Kail [Rhoades] mine; and he told me he had found it. I believe this story and I am going to let Bing Binggeli tell you his story, and I am sure it is true because I believe it.[27]

Wilford Binggeli then wrote:

> About 1931 I was working for the forest service out in the Uintah Mountains by Iron Mountain; and we were camped at Iron Mine Lake and we had been bugging all that country in there.
>
> This time we were bugging, we came around a place there and I went through some timber and small trees, and as I went through there I went out into a little opening and the ground gave away with me. It was snowing and storming, and I just luckily happened to grab hold of quaking aspen tree or otherwise I would have gone down the shaft. This man, Bill Nunley, was with me.
>
> It was quite deep, I threw a boulder down in there and it sounded like you could hear it hit two or three times. It was quite deep, 50, 75, or even 100 feet deep, I don't know how deep, but it was deep.
>
> There were foot holes in the . . . wall where you could put your feet to go down, but when I stepped on this there was an old rotten log which gave away under the ground. It was about 30 inches in diameter. I took [the] axe when I finally pulled myself up and got Bill Nunley back to where I was, he wondered what was wrong, and we opened it up and chopped it without [the] axe, and looked into it.
>
> Finally we got to looking and we could see the vein. There were several kinds of ore in this vein, but right in the center was quite a vein of real hard quartz, we couldn't even but it with an axe. Next to this quartz was a white sugar quartz, and the sugar quartz was full of gold. I got handfuls, chopped it out with my hands, and I filled both of my pockets with this stuff. It was snowing so hard we could hardly see.
>
> He [Nunley] said, "let's get out of here, you have found [Rhoades'] gold mine."
>
> This man I am talking about is an old man and he has prospected all of his life. He knew ore just by looking at it, he couldn't tell how much was in it, but he could tell you anything about ore. He saw what I had found and went practically insane and he said, "Let's get out of here.

The Indians will kill us, this is something you shouldn't find or have anything to do with."

I said, "Wait, that's crazy, let's drop some trees over this shaft so somebody else won't fall into it. Let's get our directions of where we are."

Well anyway we figured we knew just exactly where we were, so we finally got it covered over and started up the hill, and I had both pockets full. We finally caught up with the other men that were working on the ridge there and they were burning some timber when we got there. The minute we got there they said let's get out of here, so we started towards camp.

When we got to the camp, the rangers were there and told us all to get out. Well there was already about 1 1/2 feet of snow then. The rangers helped us put the chains on then they got out, and then we started to get ready and get out camps and things loaded.

When we started out, it was getting dark and snowing so hard you could hardly see anything. So we finally got started to going up the road and we even had to fell trees along the side of the road to keep the truck from going down over the side of the mountain. When we got out of there, it was the next morning.

I went to my jumper when we got down from the top because we were all soaked. My jumper was gone and I couldn't imagine where it had gone to. I looked in the truck and through everything and I couldn't find anything, then I wondered if maybe someone had taken it. I knew this man [Nunley] was honest, he wouldn't do that, so I guess I just lost it in the night.

We left Iron Mine Lake and went around Iron Mountain and down through Soapstone, and over the top down . . . and it took us all night to get out of there. From there we went on down to the ranger station and then on to Little South Fork to dry out. We knew just exactly where to go to go back to that mine.

The depression started and you couldn't get gas to go anywhere and there was no money. It was two years before we got back up in there, and we knew right exactly where to go, so we took our picks and shovels, our sacks to sack the gold, carbide lights, and candles to test the air, ropes, everything we needed to go to that mine. We thought we knew just exactly where it was, but when we got there we couldn't find it.

When Bill Nunley saw what I had found he said, "Don't tell a soul, don't tell anybody what we have found here."

He was an old prospector; he had mines down in Southern Utah. He knew ore when he saw it, he could tell you what was in it, he might not be able to tell you how much was in it, but he could tell you every kind of ore by looking at it. He was an honest man and he was a real old, good prospector.

There were a lot of people after we came home who tried to pump him to find out all about this mine. He told me never to tell anybody. He wouldn't tell anybody nothing about it. The only ones he would talk about it to were me and my son. My son and I spent $1,000 for gas trying to find this mine. Mr. Nunley and I were up in there time and time again, and we just never could find this mine. I think Mr. Powell has found something now.

When we went down and was drying out, we were down almost practically on the Stuart's ranch, they call that the Little South Fork. There was where I tried to figure out what ever happened to that gold. I got my jumper and things but could never figure out where the gold went.

We used to bank up in Kamas, and we were up there one day in the drug store; and this old man, it has been about twenty years ago, was in there and we asked him if he had ever heard of a gold mine up in that country. He says, "You bet I have heard of a gold mine." I said, "Did you ever hear of [Rhoades] coming down through this country and living here in Kamas." He says, "Oh yes, I have seen [Rhoades] come back with pack horses loaded with gold. I have handled the gold, and I have talked to him and I know where he lived. I have seen him leave here and I have seen him come back. I have seen him go up Beaver Creek up towards Iron Mountain." That was the only thing he could tell me about where he went, but he always came back down from Beaver Creek.

Then there was another person (my wife runs a beauty shop). This lady owns some stores out in Kamas, Vernal, Duchesne and Roosevelt. Her name was Mrs. Cole (she was a customer of my wife's beauty shop). I talked to her about this and asked her if she knew anything about this. She said, "You bet, I know all about that. We had an Indian girl that was working for my husband years ago, and she was going to take my husband up and show him where this mine was, but the Indian girl never made it. She was killed on the way up there."

I have checked and read and studied all of these things and it all points to where we found this mine, but I just can't find it. But I think Mr. Powell is close to it. This place that we were camped is about 1,000 feet down from Iron Mine Lake and I think that mine was up east from where the camp was. It was east of Iron Mine Lake, and I don't know how far from the camp it was, but it was in that area somewhere. We were bugging timber up in that territory, and you can still go up there and see where we had bugged and burned timber. One day I was out there hunting around on the ground and I ran into Mr. Powell. He asked me what I was doing there and I said I was looking for the [Rhoades'] gold mine that we had found once. So we were kind of working together looking around. We looked up on the other side of the creek and up on the mountain; we found a tree there with a Spanish insignia on it, an arrow. The arrow pointed right straight to where he has found this discovery now.

My son worked for the gas company as the boss at one time. He had an Indian boy that was working for him and he asked him one day if he knew anything about [Rhoades'] gold mine or any of the mines that are up in that country. He said, "You bet I do, my grandmother used to live to the side of the road about Hanna in a dugout. My grandmother told me several times Mr. [Rhoades] came down from there several times with his mules, and pack horses, and would stop and have lunch with her. She had seen the gold he had brought down with him from the hills."

Then there was another lady, I can't tell you her name, but she is an Indian woman, and I was talking to her and she said, "Oh yes, we know all about that mine." She was telling me her mother's grandfather was taken to that mine by the Indians, was blindfolded when they took him there, and they showed him the gold, then blindfolded him and brought him back out of there.

Anyway, my wife would go with me sometimes up in there. So my wife would wait for me down in the canyon and I would go up around and come back down through these places where I was looking and find her again, then we would do it over again. But anyway, several times when I was up in there I would get an awful scary feeling, I don't know what it was, but one day I was coming down through a place I just felt like I had to get out of there. I had the most awful scariest feeling, but I don't know what it was.

Another thing I would like to talk about this is if I had been alone when I found this mine, I would think I might have had a dream or it was my imagination, but Bill Nunley was with me, and he is an honest man and I trust him. It's the truth so help me, everything. Bill Nunley is dead now, he died, I don't know, I guess about 10 years ago, but he was an old time prospector and he knew what prospecting was and knew what metal was. I'd stake my life on everything I have said, and it is true about me falling into this old shaft, and about those old rotten timbers that gave away with me when I stepped on them. If it hadn't been for that quaking aspen, I guess I wouldn't be here telling about it. When he saw what I had found, he just went crazy. The gold was in the center of

> this vein and probably more gold in the other pieces of vein, and there was quite a vein or solid quartz, and right in the footwall of this shaft was foot steps going down that you could put your feet in to go down the shaft, it was storming too bad, but I did throw a rock in the shaft to see how deep it was. Then we covered it over. This vein that I looked at was about 4 feet wide, but quartz in the center, real hard quartz that couldn't break with an axe, but I did chop out some of the sugar quartz that was next to it, and then it had a kind of brown stuff that was in there too. I filled my pockets, both of them, with this ore. When we left this here mine, it was snowing so bad you couldn't hardly see where you were going. When we got up on the top and caught up with the others that were burning a bunch of timber, they said we had better get out of here and get out fast. So we started to making it back to camp, and we could hardly figure out how to get to the camp, but we just kept going and finally we decided it was right straight down the mountain from where we were.[28]

After Binggeli had recorded his story, Barney Powell continued:

> I am over to Heber at Center Creek where Bing lives, and I brought this tape recorder over to get his story. I think he has given you a wonderful story, I think he has told you the truth, and I believe it every bit of it; and I wanted to get it because I wanted to show other people what it is about, and since we have discovered this tunnel. It is about 20 feet down under the ground, and it was only through inspiration and the help of the Lord that we found this. We found air coming out of the rocks at first, we had a metal detector and we went out and camped over on the wood cutting road and walked about 1,500 to 2,000 feet west, then I had the feeling and inspiration to go down this little kind of a swale. We went down working the metal detector, and finally Dave Horn said he had a gold reading. What I mean about a metal detector, it has a red light and a green light. When the red light goes on it is Iron, and when the green light goes on it is gold. It is made especially for detecting gold, and he went around an area over 100 feet around, and when we went outside of that area the green light went out. I tell you what I believe and I know I am right that when the Spaniards—we don't know where the end of the tunnel is, we know it goes way south, somewhere south towards Iron Mine Creek is another opening that is open right now and the air goes in there and comes out right here where we are. One of the guys in our party, there were 7 of us, sat down on a pile of rock just east of this brown place and he found air coming out. "Hey, there is could air coming out of these rocks." We threw dusty dirt over the opening and

> the air coming out blew that dust up in the air. So that is all we needed, everybody pretty near went crazy then. They wanted to dig it out, and we dug it down quite a ways and we finally got down to where we could see the marks of the quartz about 2 to 3 feet think, and on the quartz on the side were drill marks, four drill holes that had been drilled from inside the tunnel up into this quartz. They had to drill that to break the quartz because it was hard. There are several pieces of quartz that show drill marks, so anyway we went ahead and dug this out to where we could see the hole. We put a 4 foot bar down in the hole and the guy put his arm down with it and still couldn't reach the bottom of it; but he could put his head down in it and he could see at least 30 or 40 feet each way north and south inside the tunnel and it was black. I have pieces of the black soot. We know that the Spaniards at that time didn't have anything but oil torches, they were not very good light and they couldn't see very good, but that is the way they mined. About every, I don't know how far, maybe 500 or 600, or 1,000 feet they would make a hole to the top of the ground. Now that is the reason we found this gold.
>
> So believe what you want to, but I know it is true, and I pray that the lord will help me. This is November 3, 1981, on a Wednesday. My name is Barney Powell.[29]

This brings us to the story of the famous lost Rhoades Mine (or should I say mines, since there were no fewer than seven worked by the Rhoadeses) that has been told and retold dozens of times in books, magazines, and newspaper articles for over one hundred years. The story has been told with bits and pieces of information, misguided truths, and embellished facts.

However, there are a few stories about the Rhoades Mines that are factual. The late Gale R. Rhoades's book, *Footprints in the Wilderness: A History of the Lost Rhoades Mines*, is the most comprehensive book on the subject.[30]

The adventures of Thomas and Caleb Rhoades span several decades and counties. Duchesne, Uintah, Wasatch, and Summit counties were places where gold was taken out by one or both of these two men. Not all the mines they removed gold from were Ute sanctioned; in fact, if the Ute Tribe ever found out that the Rhoades' were pilfering other mines, they might have met with an ugly end to their mining purposes.

As we know, time was not on Caleb Rhoades's side because he died just before the Ute Reservation was opened for prospecting. There was an agreement made on December 18, 1897, between the Uintah and White River Utes and Frederick William Hathenbruck and Caleb Rhoades that

Thomas Rhoades

Caleb Rhoades

would allow them to prospect on reservation lands for a period of ten years for a semiannual payment of $2,500, due the first day of January and July of each and every year. It must have been hard to convince the Ute Tribe to agree to such a contract, and it must have taken quite a bit of savvy on Rhoades and Hathenbruck's part. They were able to convince 114 chiefs and subchiefs of the Ute nation to sit down and hear their proposal and then agree to it. The first name on the list was that of Chief Tabby.

Others followed such as, Happy Jack, John Duncan, Jim Duncan, David Copperfield, Charley Mack, and Provo Jim, just to name a few. The main leaders of the Utes had to convince the rest to sign the contact and then live up to it, It must have been hard for some of them to do that because of what the Spanish had done to them. Provo Jim seemed to be one that tried to get along with the white people and seemed to have less trouble than most. John Duncan never trusted whites.

Chief Tabby at ninety years of age

The Rhoades Mine story is lengthy and takes extensive study to understand all that transpired during the days of Thomas and Caleb Rhoades. Many people believe they have the answers and many believe they know where the mine *is*! Do they really? I don't think they do.

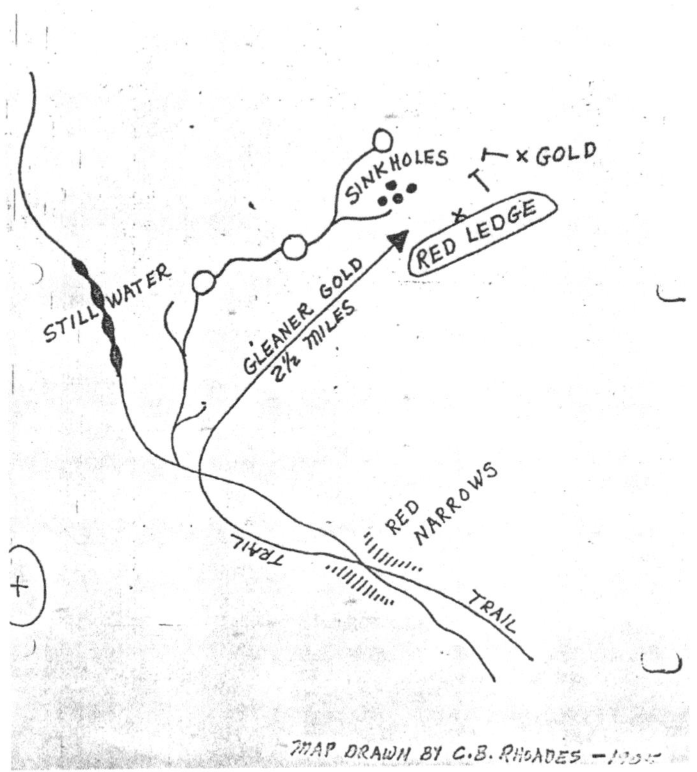

Caleb Rhoades's Gleaner gold map

THE GLEANER GOLD MINE

One of the mines that was credited as one of Caleb Rhoades's famous mines was located on Currant Creek near Fruitland, Utah. It was on one of the last maps ever drawn by Caleb before his untimely death. Titled "Gleaner Gold," he penned other clues as he drew the map for his family, hoping they would benefit from its wealth. With the death of Caleb Rhoades, this map got out; it was soon discovered that Currant Creek was the place depicted on the map. Knowing that the Old Spanish Trail came close to the area where the mine was supposed to be, prospectors headed into the hills to explore and perhaps find the mine.

LOST JOSEPHINE MINE NEAR CURRANT CREEK

In 1908, the headlines in Heber City, Utah's *Wasatch Wave* announced that the Lost Josephine Mine had been discovered near Currant Creek and that "Ancient picks and other tools have been uncovered, and ore values are sensational!" The local residents were undoubtedly excited but remembered that back in 1897 (documented in the *Wasatch Wave*) others claimed to have discovered the same mine "near the shores of a mountain lake, where the ruins of an ancient arrastra have been found." Even the values seemed quite sensational with that discovery as well.

George Olson of Heber City was one of the men who reopened the old mine near Currant Creek, and he related the story to the late author George Thompson. Olson told Thompson that after its discovery, some Provo businessmen provided the money to hire a crew of men with mining experience to open the badly caved-in mine tunnel. They mucked the old mine out to a depth of approximately 125 feet, where they made a discovery of strange and ancient-looking tools and artifacts. The mine was old and had been extensively worked many years prior. When no minable ore was found, the investors decided not to put any more money into the venture. Well, as it happens so often, greed set in, and one of the businessmen devised a plan to satisfy the investors.

A sizable sum of gold nuggets were obtained and were used to salt the old mine. The investors were brought back to the mine site so they could see what the miners had discovered. However, good judgment and guilt caused one of the businessmen to expose the attempted fraud. Now the scheme was out in the open, and nobody would have anything to do with those who had cleaned out the old mine. The mine today is caved-in and barely recognizable as a mine tunnel. It is now sitting silent in the mountains, with only an occasional hiker or wild animal that wanders by.

GOLD NEAR MOON LAKE

During Thomas Rhoades's tenure of being the *chosen one* to remove gold for the benefit of The Church of Jesus Christ of Latter-day Saints, the Utes showed him another potential mining site near the high mountain area around Moon Lake some thirty miles north of Duchesne. The site was an ancient mine, most likely dug by people more than two thousand years ago. The Spanish came up to Moon Lake in search for this ancient mine but never found it; however, they did find other mineral wealth in and around Moon Lake and its tributaries.

To qualify my remark on the theory of the mine being over two thousand years old, while interviewing Avraham Gileadi (author of several books, including *The Last Days, Types and Shadows from the Bible and the*

Book of Mormon) more than twenty years ago, he told me a strange tale that I will relate here for the first time.

At the time of the interview, Gileadi lived just below the Dream Mine near Salem, Utah, on mine property. In our discussion, we talked of the possibility of the Nephites and/or the Jaredites[31] having mined in Utah and the surrounding states. It was then that he related to me that he and a friend had accessed a special machine that could tell them where the Nephites and Jaredites mined and where they were buried. Gileadi and his friend could drive down a road and, with this machine, tell where these mines and burials were.

He told me that two Nephites were buried in the mouth of Spanish Fork Canyon on the bench not far from the big "Cross" erected there to commemorate fathers Escalante and Dominguez. Furthermore, he told me and my wife, Bonnie, that the machine directed them to Moon Lake, someplace neither men had ever been to.

Once there, they followed the direction the machine told them to go and soon arrived at a specific site. There, they found rocks all arranged in a circle under thick brush. Carefully, they removed the dirt from inside the circle of rocks and found a log base that they removed as well. Taking lights and a rope, they descended into a shaft that went down about twenty feet. Once at the bottom, they followed the tunnel westward.

Gileadi told us that what they saw next was almost unbelievable! There on the floor of the tunnel, and leaning up against the wall, were the remains of a Nephite warrior still in his armor with his weapons about him. He had mummified, but his armor and weapons seemed like they were perfect as the day this old warrior died.

Next to Gileadi, a few feet away, was a huge gold vein that encompassed the tunnel from bottom to top. It was beautiful. He told us they saw and handled other things but would not elaborate at that time. They left that place, covered it up, and marked it on a map with all the other sites they had discovered by using this special machine.

THE COPPER MAP

Much later, it was discovered that a copper map, will all the necessary markings and clues on it, had surfaced; the map would tell of the great mine at Moon Lake. The map surfaced after people like Gale Rhoades, Don Foote, myself, and a few others found clues and evidence of ancient activity in and around Moon Lake. Skeletons, cannons, and mining

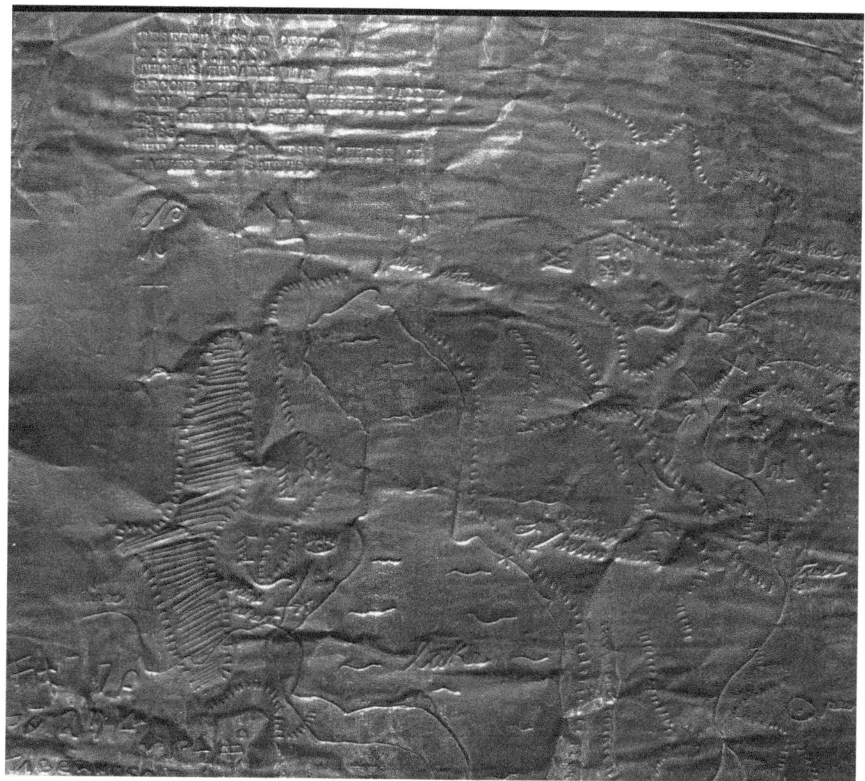

Top portion of the copper map[32]

equipment, including some worked-out mines, were discovered in Oweep Creek, Fish Creek, Duck Creek, and other places. Don Foote discovered strange inscriptions on a rock ledge not too far from Moon Lake that have since been proven to be of ancient origin. When the copper map surfaced, many thought it was an actual map made under the direction of Thomas Rhoades and Brigham Young. However, since the map is not available for testing and the present owner won't agree to let it out of his sight, we therefore have to deem it a forgery.

Whether Gileadi and his partner actually saw a vein of pure gold and a Nephite warrior we may never know. Without proof, it is only speculation and it is still just a story, as is the copper map information.

With respect to my Ute friends, I will not include sites on the reservation in Duchesne County that I have been privileged to see and to learn about. Their friendship means more to me than selling a book. So, we will leave well enough alone.

OUTLAW LOOT NEAR CLIFF STATION

Cliff Station, some twenty miles south of Myton, Utah, was, in its day, the scene of much confusion, turmoil, and excitement. The old road from Harper (where the last saloon once stood) passed by Cliff Station and came up from the south from Wellington (which is south of Price, Utah) and then came up Nine Mile Canyon north to Myton. The old road was used for many purposes, such as, moving horses, freight, outlaws, lawmen, "gandy dancers" (railroad workers), and hundreds of head of cattle. It saw its share of dignitaries as well. The saloon along this road was a common rest stop for those who had taken the grueling, bumpy ride over the rocky and, at times, muddy road.

During the heyday of the Wild West, many outlaws roamed the countryside stealing cattle, horses, payrolls, and whatever else they could get their grimy hands on. One time, five outlaws decided to heist mined gold from a group of Spanish miners who had been mining in the Uinta Mountains. The Spaniards had their gold secured on mules heading down toward the confluence of what is now the Duchesne River and Lake Fork River when they were robbed of their gold. Several Spaniards lost their lives, but the remaining Spaniards were able to hunt down most of the robbers and kill them, except for one who had left the group and set out on his own.

This robber hid up near where Cliff Station would be built years later. There, he built a rock shelter where he hid his gold. Before he had split from the rest of the group, one of the other outlaws had given this robber his part of the gold. The outlaw thought that if he was ever captured and had no gold on him he would be let off. That never happened! This robber lived in the mountains, trapping and hunting for years afterwards. He didn't go back for the gold because he never could get the horses to carry the gold. He wanted to get the gold back to St. Louis where he would be safe. As time began to run out on the old robber, he headed to St. Louis to live out the remainder of his life. It was there that he told a nephew about the gold he and the others had stolen, where and how he buried it, and about a stone map he had left that showed exactly where the cache lay buried. He figured the cache was worth hundreds of thousands of dollars. The nephew was a busy man running a prosperous freight company and couldn't take time off for a trip out west to look for a mythical treasure. The old robber, before he left the Basin, related the story to a few trusted friends, but most thought he was off his rocker. Even then, the story of a lost Spanish cache began to circulate.

Years later, long after Cliff Station was nothing but ruins, a group of rock hounds came out looking for arrowheads, pottery, or whatever the previous occupants left behind.

While metal detecting, rolling rocks over, and looking for tiny treasures, a sandstone slab was taken up that revealed some odd carvings on it. The finder took it home to Roosevelt, where it lay in his garage for a number of years, until the word came to the ears of Gale Rhoades, who asked to examine the rock since he had heard of the lost cache years before. However, the carvings on the rock meant little now because the rock had been moved from its resting place. The meanings of the etching were lost mainly because the finder didn't remember exactly where he had found the slab. However, a member of the finder's group said he would take Rhoades, his partner, and myself to the site where the slab was found. Even he was somewhat bewildered because he too had lost the exact spot of the discovery. Using metal detectors, we walked over the whole area, but to no avail. The gold is there for sure . . . but *where*?

Ruins of Cliff Station

THE SUNDANCE KID'S LIFE IN DUCHESNE

Have you ever heard of the Sundance Kid? I'm sure you have! Harry A. Longabaugh was born in 1867 in Mont Clare, Pennsylvania. He came from a very poor family of farmers who never owned land but mostly sharecropped. Harry was a wandering soul with itchy feet, so at the age of fifteen, he headed west with his cousin, George Longenbaugh (note the slightly different spelling). In 1887, Harry was convicted of stealing a horse and sentenced to eighteen months hard labor in the Sundance, Wyoming Jail. Newspaper articles said, "They laid him down in Sundance,"[33] so it was here that he earned his nickname, "The Sundance Kid." Once Harry was released from jail, he began a life of crime. In 1900, Harry became involved with Robert Leroy Parker, also known as Butch Cassidy. With Butch and the likes of Will Carver, Harvey Logan, Ben Kilpatrick, Matt Warner, and a few others, Harry became a part of the well-known gang, the Wild Bunch.[34]

Sundance Kid and his girlfriend, Etta Place

There is a theory that Harry A. Longabaugh and William Long (another man from Duchesne) were actually the same person, which would make William Long, buried in Duchesne, the Sundance Kid. The complete history of how the Sundance Kid became William Long can be found in an excellent book by Marilyn Grace and John McCullough titled, *Finding "Butch Cassidy and The Sundance Kid": Solving the "Wild Bunch" Mystery with That "Darn" DNA.*

I suggest that the reader obtains this wonderful fact-filled book to see why so many believe the William Long story. Oh, and by the way . . . there is a cache of gold coins buried by the "Kid" somewhere in Wayne County.

NOTES

1. Rufus Wood Leigh, *Five Hundred Utah Place Names: Their Origin and Significance* (Deseret News Press: Salt Lake City, 1961), 21.
2. George A. Thompson, *Faded Footprints: The Lost Rhoades Gold Mines and Other Hidden Treasures* (Dream Garden, 2006).
3. Barney Powell's personal journal, in author's possession.
4. Wilford Binggeli, Heber City, Utah.
5. For consistency, all instances of variations on the surname "Rhodes" in Barney Powell's story have been changed to "Rhoades."
6. Thomas Rhoades went to California to mine gold for the Church so coinage could be manufactured. See "Thomas Rhoades Company (1849)," *LDS.org*, accessed April 19, 2017, history.lds.org/overlandtravel/companies/403/thomas-rhoades-company.
7. On January 24, 1848, James Wilson Marshall found gold flakes in the American River at the base of the Sierra Nevada Mountains near Coloma, California. At that time, Marshall was working to build a water-powered sawmill owned by John Sutter. Marshall worked with members of the Mormon Battalion to build Sutter's sawmill. See *Wikipedia*, s.v. "James W. Marshall," last modified February 15, 2017, en.wikipedia.org/wiki/James_W._Marshall.
8. Brigham Young, when entering the Salt Lake Valley, said the following words: "This is the right place, drive on." See Richard D. Poll, "Dealing With Dissonance: Myths, Documents, and Fatih," *Sunstone Magazine*, (May 1998), www.sunstonemagazine.com/wp-content/uploads/sbi/articles/065-17-21.pdf.
9. Gold from California was used for minting coins while the gold used on and in the Temple was what was known as the Rhoades' Gold brought to Brigham Young from the Uinta Mountains of Utah. See Twila Van Leer, "Rhoads, Gold Seemed to God Hand in Hand," *Deseret News*, July 2, 1996, www.deseretnews.com/article/499383/.
10. Levi Edgar Young, "The Utah Pioneers and the Indians," *The Young Woman's Journal* 24, (1918), 20–24.
11. Wakara (Walker) was a chief of the Sanpete Band under the direction of Chief Sowiette. See *Wikipedia*, s.v. "Walkara," last modified December 21, 2016, en.wikipedia.org/wiki/Walkara.
12. Wakara had no idea what the gold was to be used for at that time. See Twila Van Leer, "Rhoads, Gold Seemed to God Hand in Hand," *Deseret News*, July 2, 1996, www.deseretnews.com/article/499383/.

13. Brigham Young did not know where the gold came. The only person that knew, besides the Utes, was Thomas Rhoades. See Twila Van Leer, "Rhoads, Gold Seemed to God Hand in Hand," *Deseret News*, July 2, 1996, www.deseretnews.com/article/499383/.
14. The Utes did not think the Mormons were "white gods." Barney Powell confused this with the coming of the Spaniards in their earlier expeditions to the New World. See Barney Powell's personal journal.
15. There is no evidence that suggested the Mormons' had a desire to enslave the Utes or any Native tribes.
16. Thomas Rhoades lived in Kamas Valley for a short period of time. See Twila Van Leer, "Rhoads, Gold Seemed to God Hand in Hand," Deseret News, July 2, 1996, www.deseretnews.com/article/499383/.
17. Shingle Creek in the Western Uinta Mountains is widely thought to be where Rhoades got much of his gold. See Dale R. Bascom, *Following the Legends: A GPS Guide to Utah's Lost Mines and Hidden Treasures* (Springville, Utah: Council Press, 2007), 49.
18. Tintic, Utah.
19. Chief Mountain Sheep, not Chief Blackhawk.
20. Farm Creek does not cross the Duchesne River. Farm Creek proper is situated between White Rock Canyon and Uinta Canyon north of Roosevelt, Utah, and empties into the Uinta River. Farm Creek Road is a road from Rock Creek over Farm Creek Pass down to Hanna, Utah. The right-hand fork of Farm Creek Pass is the name given to the road over the mountain. In early spring, during the runoff, that water empties into the Duchesne River near Hanna.
21. A Caterpillar tractor.
22. Twentieth century.
23. Mountain Sheep's place or ranch was near Pigeon Water Springs on the road to Rock Creek. Location shown to author by native Utes.
24. For consistency, all instances of variations on the surname "Bingley" in Barney Powell's story have been changed to "Binggeli."
25. Marion, Utah.
26. See Twila Van Leer, "Rhoads, Gold Seemed to God Hand in Hand," *Deseret News*, July 2, 1996, www.deseretnews.com/article/499383/.
27. Barney Powell, unpublished manuscript.
28. Barney Powell's personal journal, in author's possession.
29. Ibid.

30. Gale R. Rhoades, *Footprints in the Wilderness: A History of the Lost Rhoades Mines* (Dream Garden Press, 1980).
31. Members of The Church of Jesus Christ of Latter-day Saints believe that two civilizations called the Nephites and the Jaredites lived in ancient America. The stories of these nations can be found in *The Book of Mormon: Another Testament of Jesus Christ.*
32. This map was supposedly manufactured by The Church of Jesus Christ of Latter-day Saints, under the direction of Brigham Young and Thomas Rhoades, with the aid of James Madison Barlow. The map gives details to several mines worked by both Thomas and his son, Caleb. Printed with permission.
33. Marilyn Grace and John McCullough, *Finding "Butch Cassidy and the Sundance Kid": Solving the "Wild Bunch" Mystery with That "Darn" DNA* (Story Teller Productions, 2010), 43.
34. *Wikipedia*, s.v. "Sundance Kid," last modified March 27, 2017, en.wikipedia.org/wiki/Sundance_Kid.

Emery County

"Emery County is an immense region south of Carbon County extending from the Wasatch Plateau eastward to the Green River; it was organized in 1880. Emery County was named in honor of George W. Emery, Territorial Governor 1875–1880. 'Castle County' was first proposed as the name from Castle Valley which parallels the Plateau on its east."[1]

"WILD BUNCH" CACHE AT BUCKHORN FLAT

Following the robbery of the mine payroll in Castle Gate on April 21, 1897, in which Butch Cassidy, Elzy Lay, and Bob Meeks made away with $8,000 in gold and currency,[2] one story has it that the loot was buried in Brown's Hole in Daggett County. Another story relates that the outlaws fled south down Price Canyon to Spring Canyon. Circling the town of Price, they struck the trail leading to Cleveland, Utah, where no attempt was made to stop them.

Elzy Lay

Butch Cassidy

Bob Meeks

At Peterson Spring (at the foot of Cedar Mountain, some ten miles from Cleveland, Utah), they picked up a relay of fresh horses and rode on to Buckhorn Flat, a fifty-thousand-acre plain as level as a dance floor. The money is said to have been buried here, to be recovered later by some of Cassidy's men, who would run it in relays to the gang's main hideout at Robber's Roost in Wayne County. Whether this recovery was ever made is not known. Many believe the loot is still buried on Buckhorn Flat.

THE "WILD BUNCH" AT SAN RAFAEL

Many a treasure hunter and prospector have hunted the San Rafael Reef and Swell with some success. Just recently, a stash of gold double eagles was discovered in an old brass bedstead left in a rock cabin on the Reef. Could this have been something the gang left behind? The man who found the cache left it where it was, in fear of being punished if he brought it forth.

BURIED GOLD AT GOAT SPRINGS

Another event took place during the late 1870s. Two army officers and two enlisted men who were en route to deliver a payroll of some $60,000 in gold coins stopped at a spring to water their mounts and rest the team. Their party had been trailed by an unfriendly group of Paiutes, and they knew an attack was imminent. Just before dawn the next day, they silently buried the gold coins, mounted their horses, left the wagon behind, and made a mad dash for their intended destination. They knew the gold would be burdensome, and they had a better chance of escape without it. However, just beyond the camp, they ran into the waiting Indians, and a battle ensued. The battle was tough, but the four men came out the winners. One officer, being very greedy, decided that he would go back and get the money for himself. He then planned to tell the captain at his post that the Indians had killed the other men. He shot them all, got the money, and headed for the post. Before getting to the post, he reburied the money not far from the spring and marked it. He made it to the post and reported that they had been attacked by Indians and lost the payroll to them. However, this story did not hold because the Indians didn't want the gold, they wanted the horses. The officer was tried and court-martialed and sentenced to twenty years in federal prison.

After his release, the former army officer returned to the area where he had buried the gold coins, but he never found it. In fact, he searched for it off and on for many years but it still eluded him. Some think the treasure is hidden very near Goat Springs.

HISTORY AND FOLKLORE OF THE GREEN RIVER

The mighty Green River plays an important role in the state of Utah. Its resources are many and it is used by people, wildlife, and industry for multiple reasons. It separates several counties along its path from Flaming Gorge in the north, to the Colorado River in the south. Beginning in Daggett County, it flows past Uintah, Duchesne, Carbon, Emery, and Wayne Counties, where it empties into the Colorado River.

The Green River was called "Río Buenaventura" or "Río Verde" by the Spaniards in the eighteenth century. The Green River begins its epic journey south in the Wind River Mountains of Wyoming. The Crow Indians called the river *Seeds-ke-dee Agee*, or "Prairie Hen River"; *Agee* is Crow for "river." "By this name the upper reaches of the river were known in earliest exploratory times. The Shoshone and Utes named the river *Ka'na*, their equivalent to Bitterroot, from the great abundance in its valley of this pink-flowered herb which gave them a favorite, nutritious tuber."[3]

In 1776, the Dominguez-Escalante expedition made its way into Utah and crossed the Green River at a site just north of present-day Jensen, Utah. It is believed they were following an already-established Spanish route into the Uinta Mountains that was already over one hundred years old. They named or renamed rivers and landmarks along the way; some names are still in use today.[4]

The "Green River was known to Americans in the trappers era—1820 to 1839—as the Spanish River, since there were Spanish explorers and traders along its course and because it was in Spanish then Mexico domain. These Spaniards gave the river a descriptive name: *Río Verde*, signifying 'green river.'"[5]

John C. Frémont gave his reflection on the Green River: "The refreshing appearance of the broad river, with its timbered shores and green-wooded islands, in contrast to its dry sandy plains, probably obtained for it the name Green River (*Río Verde*), which was bestowed by the Spaniards, who first came into this country in 1818."[6] (Obviously Fremont was off by decades on his time line as to when the Spaniards first came to the area.[7])

"The Old Spanish Trail crossed the Green at an old Ute crossing south of the Book Cliffs; this historic crossing was to be known after 1853 as Gunnison Crossing."[8] The Spaniards who came hunting for mineral wealth, slave trade, and other reasons used this trail to get to the Great Basin. Other Spaniards used the better-known lower route from Santa Fe to Los Angeles for more practical purposes. Both groups referred to the Green River as *Río Verde.*

> Spanish names originally applied to physiographic features were not tolerated by the Americans who were aliens operating in Spanish-Mexican domain. The names were invariably mutated into Americanized versions; thus, *Río Verde* signifying "green river" was named Green River. But, instead of leaving the correct and obvious at that, the Americans sought and conveniently found, near or far-off, a man named *Green* "for whom the river was named," they and later history writers said. Specifically, one of General Ashley's men was named *Green*, "after whom Green River is *supposed* to have been named" (Dellenbaugh). The merging of history and folklore is here exemplified.[9]

NOTES

1. Rufus Wood Leigh, *Five Hundred Utah Place Names: Their Origin and Significance* (Deseret News Press: Salt Lake City, 1961), 22.
2. Pearl Baker, "Time Table for the Wild Bunch," in *The Wild Bunch at Robbers Roost* (Lincoln: University of Nebraska Press, 1989), 6–7.
3. Rufus Wood Leigh, *Five Hundred Utah Place Names: Their Origin and Significance* (Deseret News Press: Salt Lake City, 1961), 32–33; italics in original.
4. Ibid.
5. Ibid., 33; italics in original.
6. Ibid.; italics in original.
7. John Charles Frémont and William Hemsley Emory, *The California Guide Book: Comprising Colonel Fremont's Geographical Account of Upper California; Major Emory's Overland Journey from Fort Leavenworth in Missouri to San Diego in California, Including Parts of the Arkansas del Norte and Gila Rivers; and Captain Fremont's Narrative of the Exploring Expedition to the Rocky Mountains and to Oregon and North California* (New York: D. Appleton & Co., 1849), 72.
8. Rufus Wood Leigh, *Five Hundred Utah Place Names: Their Origin and Significance* (Deseret News Press: Salt Lake City, 1961), 34.
9. Ibid.; italics in original.

Garfield County

"Garfield County, organized in 1882—being carved from eastern Iron County—extends from the crest of the Markagunt Plateau on the west to the Colorado River on the east with its north and south boundaries parallel. It was named in honor of James A. Garfield, the then recently assassinated President of the U.S."[1]

Garfield County has a rich history. It still brings thousands of tourists in every year to bathe in its history. Most people, however, come to see places like Ruby's Inn, Bryce Canyon, Boulder Mountain, Anasazi State Park, and Red Canyon. Part of the Old Spanish Trail is located in the northwest part of the county and leaves the valley to enter Bear Valley to the west. Many have no idea that more history lies in the Henry Mountains, Escalate Canyon, and other such places.

GOLD RUSH AT EAGLE CITY

> Jack Sumner and Jack Butler found gold in the Bromide Basin on Mount Ellen in the Henry Mountains in 1889 and started another gold rush. A town called Eagle City boomed and busted quickly when the gold ran out. In 1892 discoveries in the [La Sal] and Abajo mountains triggered another gold rush, followed by news of gold in the San Juan River country.
>
> Hundreds of individual prospectors panned and sluiced with great difficulty in the slickrock country with only limited success. Lack of water, except in the major rivers, made placer mining difficult, if not impossible. The Hoskaninni Company built a huge gold dredging works in Glen Canyon at the turn of the century, and the Zahn Mining

Company ran the largest placer operation on the San Juan River in the early 1900s. All these efforts, small and great, produced little gold during the heyday of the southeastern Utah gold rushes, 1883–1911.[2]

OLD MINES IN THE HENRY MOUNTAINS

Garfield County's best-known story is that of the Josephine Mine in the Henry Mountains. The story has been written about and talked about since its conception during the days of prospectors Frank Lawler and Ed Wolverton.

Lawler and Wolverton were two hermits who spent their days hunting for gold, silver, and the Josephine Mine—a lost Spanish mine that many people had either heard of or were actively hunting for. Ed Wolverton was credited for discovering many old Spanish signs and symbols, suggesting that a large and rich mine lay hidden somewhere in the mountains. He was able to find ruins of an old mill site, a slag dump, and even a few old arrastras. Some arrastras were so old that mature pinion pine trees were found growing out of their centers. One large pine was cut down by Wolverton and was reported that it had 175 annual growth rings. But as for the mine, Wolverton never found it.

Frank Lawler stayed on the mountain for over sixty years, hunting for the mine. There are records that show it to be one of the richest mines in New Spain, and the gold that came from it helped fill the royal coffers of Spain. It equals or surpasses mines such as the Tayopa, Planchas de Plata and the El Naranjal.

It is noted that there are three sites in Utah that claim to be the location of the famous Josephine Mine; however, only Hoyt Peak in Summit County is actually documented as such.

On a pack trail, leading from the Henry Mountains to the Colorado River, one faded inscription can still be read. Scratched into a sandstone wall is "1642 Ano Dom." Between the Colorado and the Bear's Ears Pass on the Old Spanish Trail west of Monticello, Utah, is the name and date, "De Julio 1661." Near the north end of the Henry Mountains is an old burial site where Indians and Spaniards fought and were buried by Jesuit Priests. That site contains the date "1777" and a Catholic cross with an illegible name that still remains.[3]

In his journals, explorer John C. Fremont mentioned that near the Henry Mountains, he found "the very bones of a pack mule, and on either side of them, a pile of gold ore from pack sacks that had long rotted away."[4]

Other explorers have reported finding mule or horse skeletons scattered on the desert floor surrounded by rusted belt buckles and pack sacks with little piles of gold ore. Many people still hunt for gold on the Henrys but few find anything worth mentioning.

In the 1770s, a party of Spaniards discovered a deposit of gold in the upper reaches of the Henry Mountains. The deposit was exceedingly rich and the party worked the site for some time, gleaning gold from the area before heading back to Santa Fe. While en route, the party was attacked by Indians, and the gold had to be abandoned so they could escape with their lives. Not all of them made it safely back to Santa Fe, but those who did told of the experience with the Indians and what happened to the vast amount of gold they had been packing. The mine they supposedly named "Josephine," hidden in the wild and inhospitable Henry Mountains, was the source of the gold.

Enter Frank Olgean and Al Hainey, who in 1900 came into the picture with a map in hand they had obtained in New Mexico. Following this map, they came upon numerous markings and codes along a dim trail, and finally arrived at "a small bench on the side of one of the smaller peaks in the Henrys, near a hidden spring." Here they found the remains of a crude smelter and began their search for the rich mine nearby that was depicted on their well-worn map.[5]

After two or three years of intense investigation, they finally gave up, gleaning only a few colored stones to show for their efforts. Thirty years later, Hainey took his box of rocks to an assayer, just out of curiosity, expecting them to be nothing more than a box of pretty rocks and therefore worthless. But a piece of phonolite recovered from the old smelter assayed at an astounding $50,000 per ton.[6] Hainey, then an old man, tried to organize another attempt at finding the mine, but died before he was able to return to the old smelter in the Henry Mountains.

"TREASURE OF THE GOLDEN JESUS" AT THE KAIPAROWITS PLATEAU

A collection of accounts concern a treasure on Utah's Fifty Mile Mountain, also known as the Kaiparowits Plateau, in Garfield County. Around 1810, a party of Spaniards came up from Mexico, heading northward, causing hate and discontent all along their route. They even robbed and destroyed missions along the way. In doing so, they accumulated a vast amount of "gold and silver coins and bars, valuable church vessels, ornaments and statues," all of which were carried on a forty-mule train.[7]

According to legend and Indian accounts, the Spaniards, after their long, difficult, and vicious journey, found themselves short on food and water. Since game was scarce, they resorted to killing and eating some of their mules. By doing this, they had no way of carrying their booty so they secreted it along the trail. By the time the Spanish soldiers entered the Escalante Canyon area, what treasure they had remaining was hidden in a cave in that canyon and the entrance was sealed.[8]

Another account leads us to believe the cache was hidden when hostilities arose between the Indians and the Spaniards. The treasure was so cumbersome that the Spaniards had to ditch it or die trying to escape from a known fate.[9]

In all the research done on this vast horde, it is certain that one of the buried objects was a three-foot-high gold statue of Jesus. Some say it was solid gold, but that seems like a stretch.[10] If it really was pure gold, it was likely smaller than three feet, or the Spaniard's mules would have had trouble carrying it. Nevertheless, this statue seems to be the catalyst of this story. The status of whether or not the Spanish soldiers ever returned to claim their loot is unknown, but it is doubtful they did. The large horde became known as the "Treasure of the Golden Jesus," and it has tingled the minds of treasure hunters for years. Some claim that local Indians have told them old stories about a cave where "Spanish soldiers had hidden a large cross of gold and claimed to have actually seen the cave. Mysterious markings can still be found on Fifty Mile Mountain and numerous Spanish artifacts have been recovered to lend credence to the story."[11]

Yet another version of this story states that in 1875, an old Navajo man related that a huge golden image, so heavy that it took seven men to lift it, was hidden on Fifty Mile Mountain. He claimed that because of this statue, the Navajo and the Hopi have clashed as to the mountain's ownership.

To this day, this storied treasure remains hidden. But where?

GOLD DEPOSIT NEAR ROCK SPRINGS BENCH

Not too far from Rock Springs Bench, at the southern border of Garfield County, a story emerged claiming that during the late 1800s, a prospector from Escalante by the name of Tine Gray found a rich deposit of gold nuggets at a place he referred to as Davis Canyon. Davis Canyon was more than likely a name the locals used or a name that Gray adopted on his own because there is no Davis Canyon in Garfield County.

According to legend, Gray had gathered 650 pounds of gold and was in the process of packing it out when his mule up and died on him. He figured the animal must have eaten locoweed that was in the immediate vicinity. Now that Gray had no way to get his gold to market, he had to make certain that his gold would be safe while he mustered up another mule or two. He knew of an ancestral Pueblo cliff dwelling nearby with numerous clay pots inside that he could put his nuggets in for safekeeping. Hauling down several pots, Gray half buried them, filled them with his nuggets, and then completely covered them with sand and brush.

Without his mule and burdened with some necessities, he began his hike out of the rough country, searching for civilization and another mule. Gray stopped by a water hole that he was familiar with only to find it dry. Now with no food or water, he began to wander aimlessly, hunting for another water hole. It was days later that Gray finally wandered onto a ranch, where he found some help for his parched body. Gray was no young man and his seventy years of life had taken a toll on him. Gray was in such bad condition that he knew he was going to die, so he decided to share the location of his buried gold nuggets with the rancher who had taken him in. Should Gray be lucky enough to survive, he told the rancher that he would cut him in as a 50 percent partner for his generosity, but if he died the rancher could have it all. However, Gray never recovered and soon died. The rancher, knowing what to look for, set out for this mysterious "Davis Canyon" to hunt for the gold nuggets, but to his dismay, he never found them.

TREASURE AT PINE CREEK AND WIDE HOLLOW

The following is an excerpt of a letter by former miner LaVar Shurtz. He tells of a treasure that he and his father found at Pine Creek. LaVar wrote:

> When my father was a young boy he was herding cattle out of Escalante, down in Pine Creek, where the creek came out of the mountain and entered the foot hills. Right where the creek came out and the wash widened into a flat he found a large rock which he crawled up on to set down and rest. As he sat there he began pulling some of the loose rock off, (or picked up some loose rock near the large rock) and he noticed what appeared to be gold pyrite in the rock—in rock which looked like

> cement and gravel, mixed (conglomerate). He didn't think too much of the rock but took several samples home with him. Later some old man saw the rock and got all excited—said that "pyrite" was gold!
>
> Dad and I went in there years later and took samples out of the gravel bed there and the assays showed $6,000 in tellurium gold and $7,000 in platinum, just around the hill from the creek we found a large deposit of blue clay and this proved to be even richer.[12]

Shurtz also noted in his letter that his father had discovered platinum in Wide Hollow, west of Pine Creek. He feared that Wide Hollow Reservoir had covered the site where the platinum was discovered.

Not far from where Shurtz noted "Blue Clay?" twice on his map, a cross carved in sandstone was discovered. The sandstone piece with the carved cross had apparently broken loose from the cliff face (where it had been originally) and had tumbled down slope undamaged. Nearby, there were carved Pinyon pine and Juniper trees that led up into Box-Death Hollow where, when using a sluice box, fine gold can be recovered. It is apparent that the Spanish knew of this area and marked it. Markings like this were only done when the area was worth working.

LaVar Shurtz's map

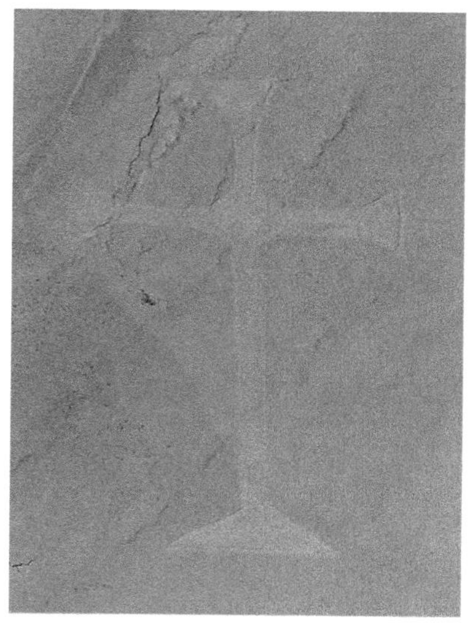

Pine Creek cross

Pine Creek marker

NOTES

1. Rufus Wood Leigh, *Five Hundred Utah Place Names: Their Origin and Significance* (Deseret News Press: Salt Lake City, 1961), 27.
2. "Minor Gold Rushes, Major Gold Production," Miriam B. Murphy, *Utah History to Go*, accessed April 7, 2017, historytogo.utah.gov/utah_chapters/mining_and_railroads/minorgoldrushesmajorgoldproduction.html.
3. "Lost Mine of the Henry Mountains," Robby Cat, *Real Treasure Hunts: Cryptic Treasures*, accessed April 7, 2017, www.realtreasurehunts.net/2010/10/lost-mine-of-henry-mountains.html.
4. Ibid.
5. Eugene L. Conrotto, *Lost Gold and Silver Mines of the Southwest* (Mineola, New York: Dover Publications, 1991), 45, 47.
6. Ibid., 47.
7. "Treasure of the Golden Jesus," Floyd Mann, *The Treasures of Utah*, forum, accessed April 7, 2017, thetreasuresofutah.yuku.com/topic/987/TREASURE-OF-THE-GOLDEN-JESUS#.WOf3sFMrKis.
8. Ibid.
9. Ibid.
10. Ibid.
11. Ibid.
12. LaVer Shurtz letter, in author's possession.

Grand County

"Grand County is a large domain extending southward from Uintah County to San Juan, and eastward from the Green River to the Colorado state line. Grand County was organized in 1890; it derived its name from the Grand River [an old name for today's Colorado River]."[1]

"The Colorado River, which runs diagonally from the Colorado [state] line through the southeast area of the County," eventually joins with the Green River in San Juan County. Prior to 1921, the Colorado River was known as the Grand River. The name "Colorado River" originally referred only to the waterway south of the junction of the Green and Grand rivers. "The Grand River apparently derived its name from the preceding Spanish form *Río Grande del Norte*; it has its source high on the western slope of the Rocky Mountains."[2]

HISTORICAL ARTIFACTS IN SEGO CANYON

Grand County has its fair share of historical outlaws, Spaniards, and lost loot. The area around Sego is interesting. This town, or what's left of it, was founded in the 1890s for coal mining. At its peak, as many as 150 people lived there. Coal seams are still burning far underground and will continue to burn for eons of time. To find Sego, travel north from Thompson, just off I-70 along the Green River. Inscriptions left by Indians, Spaniards, and more modern folks can be found along the route. Coins, gold, and other artifacts have been discovered in Sego Canyon.

FABLED GOLD CACHE AT SAGER

East of Thompson, on Sagers Flat, are the remains of an old rail stop. Here, the town of Sager flourished. Money poured in from railroad workers, cowboys, merchants, saloon keepers, and the like. Many Chinese, Japanese, and Mexican laborers were employed by the railroad company. A Japanese man employed as a Sager cook was known to have "saved and secreted all his wages. When he refused to reveal the whereabouts of his $10,000 in silver coin hoard to two ruffian railroad workers, he was killed." The two workers hightailed it out of Sager but were captured a short time later. The two men claimed they had searched the Japanese man's cabin but found nothing of his cache. It was believed by some that he had hidden his gold coins *near* his cabin but not *in* it.[3]

HIDDEN LOOT AT THE COLORADO RIVER

In or around 1906, a Denver & Rio Grande Western Railroad train was stopped near Grand Valley, Colorado. A chest of gold coins was stolen from the train during a daring robbery. The outlaws headed west to Utah, following the Colorado River to an old bridge that crossed the river. There, they attempted to open the chest, but found that they couldn't open it without destroying its contents. They buried the chest near the bridge at a depth of three feet and near three large cottonwood trees. The outlaws then continued their escape, intending to recover the cache at a later date. In due time, a posse caught up with the robbers and all were killed in the battle. Just before he died, one of the robbers told a posse member where the cache lay hidden. A search was made, but nothing was found. Subsequence searches have been made since, but again . . . nothing!

GOLD DEPOSIT ALONG THE DOLORES RIVER

Fine pieces of gold, gold flakes, and gold dust can still be found along the Colorado River tributaries, but none like that of the Dolores River. In 1856, a party of miners, returning from the California gold fields, were attacked and massacred by Indians near the confluence of the Colorado and Dolores Rivers. It was said that the miners had found a rich deposit nearby, and they chose to camp at the site on their way home to Grand Junction, Colorado. What were they doing that far south if their destination was Grand Junction? Perhaps they knew more than we know. We do know that the gold they brought back from California and the gold they had discovered at the confluence was never found.

JACK WRIGHT'S LEDGE OF GOLD

Outlaw gang member Jack Wright claimed that he found a rich ledge of rose quartz somewhere along the Horse Thief Trail, where an old Indian trail comes from the southwest to meet the Old Spanish Trail from Santa Fe to Monterrey. The ledge was supposedly somewhere between Peter's Point and Old La Sal, in a low outcropping close to the trail. A sample recovered by Wright assayed an impressive $70,000 to the ton; but, not being a miner, he was unable to relocate the site.[4] The last time anyone heard from him was just before he went off to look for the ledge one more time. He was never heard from again. Once in a while, rose quartz can be found in the vicinity, but nothing as rich as Jack Wright's ledge of gold.

NOTES

1. Rufus Wood Leigh, *Five Hundred Utah Place Names: Their Origin and Significance* (Deseret News Press: Salt Lake City, 1961), 29; italics in original.
2. Ibid., 29–30.
3. "Sager Silver Cache," Floyd Mann, *The Treasures of Utah,* forum, accessed April 10, 2017, thetreasuresofutah.yuku.com/topic/982/SAGER-SILVER-CACHE#.WOwLRFMrKis.
4. Ron, "Utah Red Gold," *Treasure Hunting,* okietreasurehunter.blogspot.com/2009/02/utah-red-gold.html.

Iron County

"Iron County was one of the original eight in [the] Utah Territory. Prior to the Act of Congress creating the Territory in 1850, the area now embraced in Iron County together with an almost boundless expanse across the Colorado Plateaus on the east and to the Sierra Nevada on the west, was known as 'Little Salt Lake County,' thus named from Little Salt Lake in the sink of the Parowan Valley. Reconnaissance of the region in 1849 disclosed iron ore in the Iron Range to the southwest; in consequence of this discovery, plans were made to colonize the Little Salt Lake Valley; and Iron County was established by decree as the first step for colonization. The County was thus named from the large bodies of ores within its bounds, on the west of Cedar Valley."[1]

A mission was organized under the leadership of The Church of Jesus Christ of Latter-day Saints under the direction of George A. Smith, who was to colonize the new county and attempt to smelt iron ore. This LDS mission comprised of 119 men, 310 women, and 18 children, who came from Salt Lake City on December 7, 1850, and arrived at the gathering site of Parowan, January 13, 1851. Parowan was the first settlement south of Utah Valley on the Mormon Trail, which actually followed much of the Old Spanish Trail.[2]

The ghost town of Iron City is preserved for adventurers and historians alike. The site includes "the ruins of an ambitious pioneer industry, the [settlers'] second attempt to smelt iron ores at their source." The ghost town is on the "banks of Little Pinto near the south base of Iron Mountain . . . a rich mass of magnetite and hematite ores in the Iron Range. The ruins comprise a well-preserved mammoth charcoal oven, blast furnace, primitive stone-lined pit for grinding the ores, and a pretentious residence. Since coking coal had not been found to use in smelting the iron ore, large quantities of timber were reduced to charcoal in beehive-shaped ovens. Lack of coking coal and with the railroad bringing in iron products from the East after 1869, this pioneer attempt to smelt iron ores was abandoned." Seventy-five years later, Iron Mountain would come alive again when the steel giant, Geneva Steel, began transporting iron ore from the mountain to its furnaces in Utah County.[3]

LEGENDS OF HIDDEN GOLD AT MOUNTAIN MEADOWS

There was nothing so daunting, or so repulsive, as the massacre at Mountain Meadows, which at the time was located in Iron County. Iron County will forever be remembered for this tragic event. Much has been written over the years, including a detailed account of the massacre by author and historian Juanita Brooks in her book, *The Mountain Meadows Massacre.* Brooks captured the agony, spirit, and damning effects this event had on the survivors of the massacre, including the fear of death that encompassed the members of the California-bound immigrant wagon train known as the Baker-Fancher Party.

Juanita Brooks

Over the years, tales have been in circulation in regards to a supposedly large cache of gold coins buried by the party before the massacre. Whether this is true or not, it still brings the treasure hunter and their metal detector to Mountain Meadows hoping to find this lost loot.

The basic account, current for many years now, essentially maintains that the Paiutes initially attacked the wagon train (most likely under urging or encouragement from local Mormon leaders), but the emigrants were able to repel the attackers after some loss of life and injury. Since the situation took place in the emotionally charged times that followed the 1856 Mormon Reformation, and with the prospect of a war with federal troops looming in the horizon, the Mormons were more likely to become involved with the emigrants, who they perceived to be antagonists.

John D. Lee sitting on his coffin

However, history tells us that the local militia (mostly Mormons, led by local church members John D. Lee and Isaac Haight) approached the besieged wagon train under a flag of truce. They convinced the emigrants to surrender their weapons, promising, in return, a safe escort out of the area. The desperate emigrants agreed, only to be slaughtered by their would-be protectors after only a few yards from the safety of their wagons.[4]

John D. Lee

Interviews with Paiute historians call into question the events surrounding their involvement in the massacre. There were problems with collecting of this information due to the lack of firsthand knowledge and the language barrier between the Paiute storyteller (historian) and the interviewers who were white.

Clifford Jake of the Shivwits Band of Paiute related the following to me in 1998:

> I used to chop wood for the old man Isaac Hunkup and his sister. . . . He was telling me a story, telling me what they see and what they hear also. And the Mountain Meadow massacre and Paiute didn't know anything about what was taking place over there. They were calm and quiet. They didn't know nothing about nothing.
>
> There was two brothers that come to the pine valley, hunting deer. . . . But what he was telling me was that they were there camping out there in the mountain. In the morning during the day [they] heard a gun, like popping, popping like a firecracker. So they went up on the mountain. There was a wagon train the people where people were shooting and killing the wagon train people, is the way he used to tell it. Oh, my goodness!
>
> Two guys were still waiting when they got down, they got everything, everything. Even their horses, the wagons were tipped over, they had some cows and sheep and the pigs and chickens and the womenfolks

also. They got womenfolks. They were running around and getting shot there. They were watching from a knoll. Them two guys.

"Oh, my . . . ," they said, "they are killing them people." . . .

No Indians live around this area. This is their hunting place, not the pine valley.

So, anyway, they got down, they got all of those things. Those things they took away from the settlers, the wagon train. And they talked together. "Let's follow the rim about a mile, a couple of miles, away from them, see what they are going to do." So they went.

They took all of them people that [were] massacring the wagon train. They went over towards the east. They followed them quite a ways from they followed them till they get to the place to where they are going to change their clothes. So, anyways, they followed them clear to New Harmony. From there they sneak up on them about a half mile. They watch them and they watch them. They sit there. They clean theirselves; they took off their Indian outfits off—clothes, Indian clothes. And they were white people. Them white people, they washed themselves up and cleaned themselves. They were white people that done it.

And they said, "Let's get going," they said. "Let's get going to warn them other people down to Sham [the Paiute encampment]."

They traveled to get there as fast [as] they can. I don't know if they were on a foot or on a horse. But, anyway, they made it down there . . . to get a hold of them Indians, house to house. I want them to be aware. We are going to [be] blamed for something that we didn't have happen. For those people, for shooting them wagon train. Better beware.

Poor Paiutes

They said they got really scared. After awhile during that day one of the guys from the younger Indians they saddle up their horse and warn the people around the area. Clear to Cedar City and . . . maybe Moapa too. So beware; we are going to get blamed, going to get blamed for what those white people did. There were no Indians in that massacre. . . .

> The authority came down. They got there. They said Indians don't leave their dead like this. They started blaming the Indians for it. The Paiute Indians around this area, they didn't know anything about what happened over there. They didn't even know nothing. There weren't no Indians around that place there. . . . That's what takes a place that time. Us Paiute nation got blamed for that.[5]

About the same time that I interviewed Clifford, I also talked with a private investigator named Will Rogers, who heard an account of the massacre from a man named John Seaman. Seaman told a similar story about a group of white men, dressed as Indians, attacking the wagon train.

When Brigham Young, president of the LDS Church at the time, learned of Lee's and Haight's roles in the massacre, he excommunicated both men. In 1874, Lee was executed for his involvement in the crime.[6]

Despite the disciplinary action taken by the church, the Paiute brothers' fear was realized, and the Paiute nation carried the principal blame for the massacre for more than one hundred years. However, since then, private investigations such as the ones conducted by Juanita Brooks and myself have helped to shed more light on the event, the people involved, and their reasoning for attacking the wagon train.[7]

In 1999, Gordon B. Hinckley, then president of The Church of Jesus Christ of Latter-day Saints, "joined with descendants of the victims to dedicate a monument at the [massacre] site."[8]

In 2007, on the 150th anniversary of the Mountain Meadows Massacre, LDS apostle Henry B. Eyring issued an apology to the Paiute nation: "The Paiute people who have unjustly borne for too long the principal blame for what occurred during the massacre. Although the extent of their involvement is disputed, it is believed they would not have participated without the direction and stimulus provided by local Church leaders and members."[9]

Code on the map
(see page 86)

Because of this massacre and because it happened on the Old Spanish Trial, many a prospector, outlaw, adventurer, and more have come looking for lost or buried loot. Even some Indians talk of buried treasure hidden by both the Spaniards and the settlers moving through Mountain Meadows.

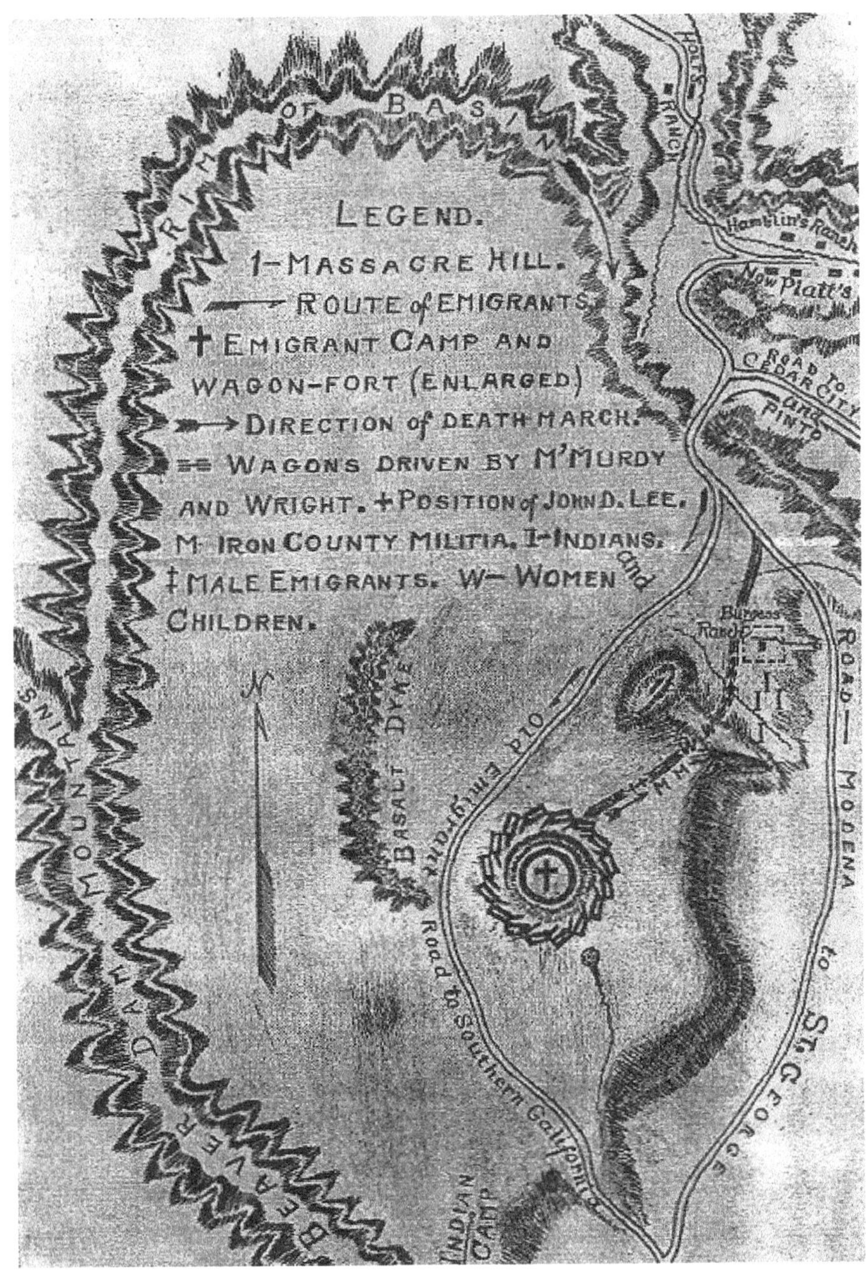

Massacre Hill

A strange map surfaced many years later that claimed a cache was buried on a knoll at Mountain Meadows. It is said that the map is in code and if you can read the code you can find the cache.

Noted on the map are the letters "W" and "W," and below is "19M7," a wavy line, and the number "4," which some believe is a code that might lead to the lost cache of the Baker-Fancher Party. Did John D. Lee know about the possibility of a stash of gold coins hidden by the frantic party before they were murdered? Perhaps, but as far as history tells us, he never let on.

GOLD NEAR WINN CANYON

One of the stories that has circulated over many years is the one of a lost gold ledge somewhere near Winn Canyon in Paragonah. The Fish Lake section of the Old Spanish Trail was followed by a group of explorers led by William Wolfskill, Lewis Burton, and George Yount. In 1830, they were traveling south, crossing the headwaters of the Sevier and Virgin Rivers to the Little Salt Lake (just northwest of present-day Paragonah), when they were besieged by a vicious, blinding snowstorm. They sought relief by finding a spot in Winn Canyon, where they built a fire and were able to keep their animals reasonably safe and comfortable. However, the storm proved to be the worst of that year and consequently some of their horses froze to death. The men had to go out further to find suitable firewood to burn and keep themselves from freezing. They ate horsemeat to stay alive, since most of their provisions were consumed during that awful storm.

At one point, while searching for firewood, the men discovered a ledge of rock that was streaked with gold. Being stranded in the mountains and far from needed supplies, they could only mark the area in hopes of someday returning to mine the gold ledge. On a black boulder near the mouth of the canyon, they inscribed a cryptic message reading "Gold 1831," with what appears to be the initials "W," "LB," and "Y"—initials of the three men mentioned above. However, after the storm passed and the men resumed their journey,

The Wolfskill gold ledge boulder

Wolfskill and the others became rich in California. They never had a need to return to Winn Canyon and the gold ledge they left behind.

In the *Iron County Advocate*, published Thursday, November 12, 1992, A. True Ott penned the following:

> As stories go, the tale of three mountain men, snowbound in a mountain canyon in Iron County while striking Spanish gold, is surely as good as any, but what are the facts?
>
> According to the "Utah Historical Journal," Lewis Burton, William Wolskill *[sic]* and George Yount, along with a string of cattle and horses and a number of Spanish hired hands, left Taos, N.M., on Sept. 30, 1830 on a beaver trapping expedition along the Old Spanish Trail.
>
> Breaking off the trail to trap the Sevier River basin, the party found themselves atop the Markagunt Plateau in a mountain blizzard. Trying to reach lower elevations, they dropped off Summit Peak by way [of Winn Canyon] where they were forced to spend the winter.
>
> Somewhere in this canyon, the mountain men found gold—enough gold that Wolskill *[sic]* and Yount purchased a vast amount of land from the Spanish at Monterrey, Calif., and eventually became very wealthy ranchers.
>
> The fact that these early mountain men found a substantial amount of gold is well documented, but did they mine it or was it already processed for them by earlier Spanish expeditions?
>
> The answer may well lie in the identities of their Spanish guides, for it is highly probable that they came along for more than just beaver furs.
>
> Carved into a black volcanic boulder [in Winn Canyon] is a detailed map or trail marker of some kind that obviously is centuries old and not Indian.
>
> Also carved into another rock further up the canyon is the word "gold" in English text, with the letters, . . . the date 1831.
>
> TW could be Wolfskill's initials, while LB could be Lewis Burton, for the dates match.
>
> Whatever occurred in [Winn Canyon in] 1831 has faded into the mists of time, with only cryptic reminders whispering of those who came before.[10]

A. True Ott made one mistake in his article when he mentioned a "detailed map or trail marker" in connection to Wolfskill's gold ledge. In fact, this marker has no bearing whatsoever to this story. However, it does relate to a different story.

SPANISH TREASURE SIGNS NEAR PAROWAN

Not far from Paragonah is the town of Parowan. Today, Parowan is known as the jumping-off spot for Brian Head Ski Resort. Before that, it was a bustling Mormon outpost. "The name Parowan has evolved from Paragoons, the name of the [Piede] band living near the Paragoon, the lake in the sink of Parowan valley. This [Paiute] word Paragoons means "marsh people."[11]

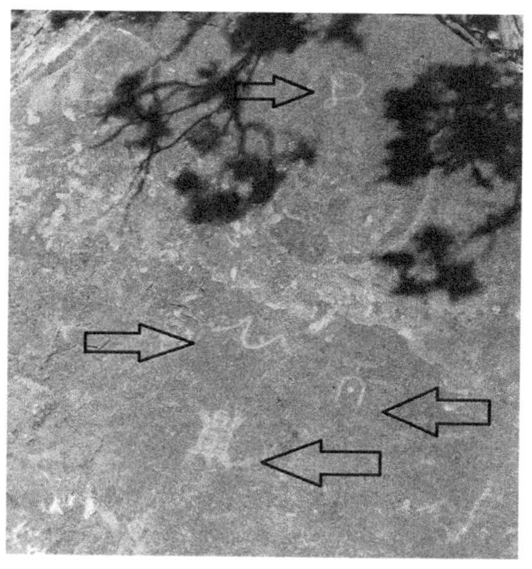

Spanish treasure signs

It is well-known that the Old Spanish Trail came through what is now Red Creek, east of Paragonah, but fewer people know that there was another route further south. Spanish symbols have been found along a well-worn trail from Parowan Canyon through Moseman Canyon and into Summit Creek. So, why leave the main trail only to converge back onto it eight miles further south? The answer may lie in the characters found carved on a gray basalt boulder not far from the Parowan Canyon road.

The Cross of Lorraine was used primarily by Spanish miners when a mine or treasure was very nearby.[12] This cross is apart from the rest of the symbols, which denotes its importance in the overall message. Noted in the image of Spanish treasure signs is the cursive letter "D," which signifies "5" or "500." This particular "D" has all the earmarks of the nineteenth-century cursive style used by the Spaniards at that time. The wavy line in the center of the photo is similar to the sign that means "Fifty varas away" (a *vara* is "a unit of length in Spanish- and Portuguese-speaking countries, varying from about 32 inches to 43 inches"[13]), but the wavy line can also indicate either water or a valley. The horseshoe with the dot in its center indicates a "mine" or "cave" with (dot) gold inside. A "mine" can be a ledge

Cross of Lorraine

of gold or silver. It doesn't have to mean an actual mine tunnel. The "turtle" sign is a dead giveaway as to the richness of the mine. The turtle is complete and it is directing the follower to the wavy line. It makes sense that this is a good location for gold. Is this what Wolfskill found in Winn Canyon?

MYSTERY GLYPHS IN CINDER CANYON

In the tiny hamlet of Summit, Utah, and further south in Cinder Canyon, or as some refer to it, Cinder Hill, are several rock outcroppings covered with Indian petroglyphs. However, some believe not all the markings were made by Indians. They believe that one glyph in particular is a treasure sign made by other ancient inhabitants that once roamed the area. They believe a vast horde of gold lays hidden in the canyon under the black cinders of an extinct volcano. It may be true that an extinct race of people once roamed here, but a treasure? One can only hope.

Hiking these mountains, one can find two important signs that are not by any means made by Spanish miners. The two signs on page 93 are actually Iberian, not Spanish. They indicate that two people, who were significant to the gravediggers, are buried there.

At the mouth of Cinder Canyon is a rock with what is called a "square circle" glyph (see below). This is the boulder that A. True Ott referenced in his article earlier, but how he came to the conclusion it was a map or trail marker is a mystery.

Mystery glyph at Cinder Canyon

This rock is only the beginning of a long line of petroglyphic signs that are found in the area. It is thought that none of these glyphs have Spanish characters on them; however, at the top and above this particular glyph, the Iberian symbols can be seen that are counterfeit to the real ones some 150 yards further south.

Iberian signs

Counterfeit symbols are identified by using a microscope camera to identify mica schists that tell us the age of the glyphs.

Between Paragonah to the north and Summit to the south, it seems as though the Spaniards, Iberians, and Wolfskill and his men may have stumbled across a land rich in gold.

IDAHO BILL AND THE STAGECOACH ROBBERY

One of southern Utah's true treasure tales concerns an outlaw named Idaho Bill: a man who plays a primary role in the story of an unsolved treasure robbery. The robbery took place at a stagecoach rest area that would be across the highway from present-day Modena. Idaho Bill and his gang terrorized mail carriers and travelers all along the road between Beaver and Pioche (a mining town just across the Utah-Nevada border), as well as on the "Hugh White" road from Silver Reef.

Tipped off by a confederate at Pioche that a northbound stage would be carrying a heavy treasure box on December 13, 1875, Idaho Bill made plans for a lone robbery. When the stage pulled into Desert Springs after dark, Idaho Bill was watching from a hiding place out in the sage and the cedars. When the half-frozen driver and his passengers went inside the station to warm up, Idaho Bill quietly threw the treasure box to the ground and dragged it off into the darkness. The loss was discovered within minutes, but no one was fool enough to follow the trail into the desert to an almost certain ambush.

The theft was reported the following day in the *Beaver Enterprise* newspaper: "The Desert Springs station agent reports that the stage was robbed there yesterday evening. While the stage was stopped for supper and to change horses, the treasure box was taken and the mail bags rifled. There was said to be valuable cargo in the box, and missing express mail is already being inquired after."

Wells Fargo immediately dispatched their crack investigator, Frank Blair, to the scene. With Sheriff Cooms and a posse, Blair searched the desert country, but they never found Idaho Bill, because he was in Beaver with Al Winn, both drunk and shooting up the town. While the two outlaws were intoxicated, Judge Boreman and the town marshal locked them in the town jail.

When Agent Blair learned that Al Winn had an airtight alibi, Winn was released, but Idaho Bill was held for robbing the mail. To Blair's disgust, Al Winn hired a shyster lawyer who got Idaho Bill released on a $3,000 bond: a large bail at that time.

Both Idaho Bill and Al Winn made a beeline for Desert Springs, where they intended to kill old Ben Bower for testifying against them, but Bower was still in Pioche, so they shot up the station and took enough whiskey to get them back to Beaver. Bill made a big mistake by returning to town, for while he was gone, Agent Blair had obtained a new arrest warrant for the stagecoach robbery.

Idaho Bill went back to jail, but he became so violent, threatening his guards and nearly tearing the building apart, that Judge Boreman ordered him into the Territorial Prison, although he hadn't even gone to trial yet. Two weeks later, he was officially sentenced to ten years. Prison records revealed that he began his sentence on March 19, 1876.[14]

What happened to the treasure he stole at Desert Springs stage stop, or to all of the other loot his gang had taken? We may never know.

A little less than two years later, on September 5, 1877, readers of the *Salt Lake Herald* read the following headlines: "Escaped. Five Dangerous Convicts Unceremoniously Leave the Penitentiary."[15]

Five desperate prisoners, including Idaho Bill, took their exit by "digging under the building and, by means of blankets torn into stripes *[sic]* scaling the outer wall."[16] They were eventually killed in a gunfight, however, and the lost Desert Springs treasure box of $2,000 was never found. It's still out there, waiting to be discovered.

CEDAR BREAKS'S LOST SILVER MINE

A rich gold outcropping in black hornblende, assaying at $5,000 to the ton, is reportedly located somewhere in the high country between Cedar City and Cedar Breaks National Monument. An old mine was located in the vicinity of Ashdown Creek by the late Paiute chief Wilford Jake back in the 1930s.

Jake had been dropped off by a buddy up at the top of the monument because Jake wanted to explore the region below. He told his companion to meet him on the highway (SR 14) down below. According to Jake, he climbed over the fence and slipped down over the edge of the cliff, dropping some twenty feet or so into soft sand.

"I went clear up to my knees and had a dang hard time getting myself out," he later told me.[17] From there, Jake wound his way down the mountain, examining everything he laid his eyes on.

Soon he came to a set of cedar poles sticking out of the ground with old rusty wires wrapped around them. He was confused as to why someone would stick four cedar poles in a six-foot square and then put wire around them. He soon found out why when he stepped over the low-hanging wire and his weight caused the ground below him to cave in. The next thing he knew, he was sitting in an old ore cart filled with rotted old bags.

Once he was able to shake off the sudden impact of falling into the ore cart, he began to assess his surroundings. He discovered that the hole he came through was an old air vent; he was in a long mine tunnel with a dozen old ore carts filled with pure silver. He took some of the silver, then climbed back out of the hole and got on his way to meet his friend.

Years later, he made the trip back to where the cedar poles had been set around the vent hole, but a flash flood had carried the poles down the mountain and the vent hole was nowhere to be seen. However, he marked the spot on a map with the hopes of someday getting the right equipment in order to relocate the old mine tunnel and the rich hornblende silver in the old ore carts.

NOTES

1. Rufus Wood Leigh, *Five Hundred Utah Place Names: Their Origin and Significance* (Deseret News Press: Salt Lake City, 1961), 42.
2. Federal Writer's Project, *The WPA Guide to Utah: The Beehive State* (Trinity University Press, 2013).
3. Rufus Wood Leigh, *Five Hundred Utah Place Names: Their Origin and Significance* (Deseret News Press: Salt Lake City, 1961), 41–42.
4. Richard E. Turley Jr., "The Mountain Meadows Massacre," *Ensign*, September 2007.
5. Clifford Jake, interview by Stephan Shaffer, 1998.
6. Richard E. Turley Jr., "The Mountain Meadows Massacre," *Ensign*, September 2007.
7. For more information, see *The Mountain Meadows Massacre* by Juanita Brooks or *Massacre at Mountain Meadows* by Ronald W. Walker, Richard E. Turley Jr., and Glen M. Leonard.
8. "Mountain Meadows Massacre," *Newsroom*, accessed April 10, 2017, www.mormonnewsroom.org/article/mountain-meadows-massacre.
9. "150th Anniversary of Mountain Meadows Massacre," *Newsroom*, accessed April 10, 2017, www.mormonnewsroom.org/article/150th-anniversary-of-mountain-meadows-massacre.
10. A. True Ott, "Treasure may not be best find," *The Iron County Advocate*, November 12, 1992, www.three-peaks.net/SpanishGold01.pdf.
11. Rufus Wood Leigh, *Five Hundred Utah Place Names: Their Origin and Significance* (Deseret News Press: Salt Lake City, 1961), 74.
12. Author's personal research.
13. *Dictionary.com*, s. v. "vara," accessed April 10, 2017, www.dictionary.com/browse/vara.
14. James B. Hume, John N. Thacker, *Wells, Fargo & Co. Stagecoach and Train Robberies, 1870–1884: The Corporate Report of 1885 with Additional Facts about the Crimes and Their Perpetrators*, ed. R. Michael Wilson (Jefferson, North Carolina, McFarland & Company, 2010), 208–9. See also "Children of the Massacre," *Olivercowdery.com*, accessed April 10, 2017, www.olivercowdery.com/smithhome/1850s/MMMbill.htm.
15. "Escaped," *Salt Lake Herald*, September 5, 1877, newspapers.lib.utah.edu/details?id=11718326.
16. Ibid.
17. Wilford Jake, interview by Stephan Shaffer, 2009.

Juab County

"Juab County, created in 1852, is in west-central Utah extending from the Wasatch Range westward to the Nevada line. The eastern fourth of its boundary is very irregular, following the crests of several short mountain ranges; west of Tintic Mountains, Juab's north-south dimension is narrow in comparison with Tooele on the north and Millard County on the south. By extension, the county was name from that of Juab Valley."[1]

"Juab Valley, in the eastern part of the county, parallels the west base of Mount Nebo and continuing mountains. . . . *Yoab* was the earlier form of the name; this was used in the Third Epistle of the Mormon [LDS] Presidency. [Historian Hubert Howe] Bancroft writes, *Yo-ab* is the name of a friendly Sampitches Ute; the word signifies "flat or level" and is in reference to the Juab Valley floor."[2]

I note here that at the confluence of the Duchesne, Green, and White Rivers in eastern Utah, the White River Utes called that plain *Yu-av*, which is "level or flat."[3]

FABLED GOLD CACHE ON SWASEY MOUNTAIN

Stories both factual and fictional can be found in the county's past. One such story emerged several years ago, concerning one young man by the name of Scott Taylor who supposedly discovered a cache of gold bars, weapons, and artifacts concealed in an overhanging ledge somewhere near or on Swasey Mountain—an extension of the House Range in southern Juab County.

Apparently Taylor contacted officials from the Bureau of Land Management about the discovery and demanded 40 percent of the find *if* he were to reveal the location to them. Of course, the BLM rejected his demands and began to browbeat Taylor by intimidation and threats, trying to get him to reveal the location to them. Suddenly, Taylor conveniently forgot where the site was and blamed the lack of memory on his prior use of drugs. The BLM kept a close eye on Taylor for some time but he did not return to the site.[4]

One thing that kept most serious researchers from believing Taylor was his description of the bars of gold. He said they were stamped by the US Calvary. In addition, Taylor claimed that Civil War artifacts were among the gold bars. Simple fact states that the US Calvary never stamped gold bars, and Taylor's claim that Civil War artifacts were cached there as well can only suggest that he assumed what he saw came from the mid-1800s, causing him to associate them with the war.

Taylor also claimed that Indian arrows and other artifacts of that nature were cached. If what he said is true, it sounds more like an Indian cache site with artifacts from several raids by them. As far as the bars of gold . . . it's anybody's guess at this point.

GOLD CACHE AT DUGWAY PASS

"Two miles west of the old Dugway Stage Station, a party of four gold miners were killed at Dugway Pass by an Indian war party. Years later, one of the Indians stated that all of the miner's belongings, including several packs filled with gold, were thrown into deep cracks in the rocky ridge above the pass."[5] In 1975, a rock hound discovered a rusted revolver and gold pan near some rocky outcrops, just north of Dugway Pass.

EMERALDS ON DESERT MOUNTAIN

In 1976, a man by the name of Darrell Hendricks was prospecting near Desert Mountain when he met a young woman and a middle-aged man wandering nearby. Striking up a conversation with them, he gleaned the following story. The woman hailed from the east coast and the gentleman was her lawyer from Chicago. Together, they had set out to locate her father and his lost mine. She claimed that, through the years, her father would leave his home in the east and travel to Utah, where he would spend a month returning home with enough money to keep the family in good order for the rest of the year. The following year, he would leave again to repeat the process. She claimed that what he had discovered was not gold or silver, but emeralds. The last time she saw him, she asked him to draw a map of this deposit of emeralds. She told Hendricks the reason behind her request was because her father was getting on in years and she wanted the information so she could carry on his work.

After several weeks, and not hearing from her father, she contacted an attorney she knew for help. With map in hand, the two headed for Utah and the vast west desert to find her father. However, after many days searching, asking questions, and getting no answers, they finally ran into Hendricks. He told them that he knew of gold and silver deposits, but never had he heard of emeralds in the district they were searching. Alice (name changed) showed Hendricks the hand-drawn map her father had given her. He knew several of the landmarks, but there were other marks on the map that he didn't understand. After spending the day with them, Hendricks left them his contact information and left for Salt Lake City and home.

Two years later, a letter came to Hendricks with a map inside. It was the map to the emerald deposit. Apparently, the daughter of the old prospector had given up trying to find her father or the emeralds and, having nobody to give the map to, figured Mr. Hendricks would benefit more than she. The letter included the following request: *In any case Darrell if by chance you ever find the emeralds I would be very appreciative if you would disclose that information to me as I feel my dad would have been close by.*

Somewhere around Desert Mountain, west of the Little Sahara sand dunes, the emerald deposit lies hidden amongst the gray boulders and barren landscape. Perhaps the old man died there, digging for his precious emeralds.

NOTES

1. Rufus Wood Leigh, *Five Hundred Utah Place Names: Their Origin and Significance* (Deseret News Press: Salt Lake City, 1961), 43.
2. Ibid.; italics in original.
3. Ibid.; italics in original.
4. Travis Reed, "Man says he's found treasure, will let it sit hidden in desert," *Arizona Daily Sun*, June 17, 2005, azdailysun.com/man-says-he-s-found-treasure-will-let-it-sit/article_03a3bc7b-b065-5503-8e09-acbcff864af7.html.
5. "Juab County," Floyd Mann, *The Treasures of Utah*, forum, accessed April, 10, 2017, thetreasuresofutah.yuku.com/topic/972/JUAB-COUNTY#.WOwIm1MrKis.

Kane County

"Kane County is central to the tier of three southern counties. Most of its area is south of the plateau country and is of comparatively low altitude; nearly all is in the Colorado River drainage. Kane County was named in honor of Colonel Thomas L. Kane of Philadelphia, a sincere friend of the Mormon people [LDS] and through whose prestige in Washington the stress of the "Mormon War" [Utah War] was relieved. A heroic bronze statue of Colonel Kane was placed in the rotunda of [the] Utah State Capitol in 1959."[1]

MONTEZUMA'S TREASURE CAVE

Without a doubt, the best of all stories in relation to Kane County is that of Freddie Crystal and Montezuma's Treasure Cave. The story has been told and retold over the years and the treasure has been equally hunted for without success or hope of ever finding it. The mystery still lingers, and as long as it does, people will still hunt for that elusive Aztec gold.

Montezuma

Freddie Crystal appeared in Kanab in 1914 and landed a job as an all-around handyman on a ranch owned by Oscar Robinson. On Freddie's time off, he would be seen exploring the canyons, trying to match Indian petroglyphs with characters on a map that he carried with him. Freddie confided in Robinson that he was searching for the fabulous Montezuma treasure that was taken out of Mexico to keep it from falling into the hands of the Spanish conqueror, Cortez. He had spent years searching for the matching petroglyphs all over the southwest and the Arizona Strip, and as soon as he found them, he would have the treasure.

Then one day, Freddie disappeared without so much as a farewell to his boss Oscar Robinson. Where he went nobody knew, and they assumed it was the last they would see of Freddie Crystal.

Then, one morning in 1920, Harriet Robinson Judd, a daughter of Oscar, went to the bunkhouse to call the hands to dinner and as she entered the room, there on a bunk sat none other than Freddie Crystal. It had been four years since he left and now he was back and acting like he had never left. The foreman of the ranch, Alvin Judd, did not want Freddie back, but old Oscar had a soft spot for Freddie and he convinced Alvin to put him back to work.

One day, Freddie got Oscar alone and showed him an old parchment that he had obtained in Mexico where, by the way, he had been the past four years. The old parchment was several hundred years old and on it were signs, symbols, and landmarks that led to where Montezuma's treasure was hidden. Oscar knew Freddie would be out hunting for this mythical treasure, so he made sure Freddie had enough time off for his adventures.

Freddie gained a friend during his time on the ranch and took him along on his excursions. On one particular day, while standing on a high point of land looking for a landmark, he suddenly yelled out to his companion that he could see the mountain that was marked on the parchment. Excitement overwhelmed Freddie to the point that he couldn't wait to get to that mountain some twenty miles distant. After traveling the twenty miles to the mountain, Freddie and his friend gazed upon a series of hand-carved steps, or holes, the size of human feet. These steps were placed precisely and led five hundred feet up to a crevice, or aerie, near the peak's summit. This ended up being a natural lookout place. The treasure would be found on the next mountain to the east that glistened chalk-white in the sun and was known as White Mountain.

After reaching White Mountain, Freddie found the spot he was looking for, as shown on his weathered piece of parchment. He quickly drew out his knife and began digging into the hard surface. It was a man-made type of concrete that Freddie described as a plug intentionally placed there.

Freddie described the plug as a concrete and blue limestone mixture, but the only blue limestone that anyone knew of was miles away on the far side of the Colorado River near Page, Arizona. Once the plug was removed, it revealed a tunnel running back into the mountain. The tunnel was sixteen feet tall and was fourteen feet wide. Freddie was excited because he thought he was at the threshold of recovering the richest treasure on earth.

Was the treasure here?

Realizing that he couldn't retrieve the treasure with only the two of them, Freddie decided to get help, and the only place he could acquire it was from the town of Kanab. At that time, the town was run by what was commonly referred to as the "Petticoat Government" since the mayor and the city council were all women. It was the first government of its kind in the United States. Once Freddie let out his little secret, the mayor and council approved a measure to help him in his attempt to get to the treasure. The people of Kanab were more than enthusiastic about the prospect of hunting for treasure, but more important what that treasure could do for the community.

The city council issued a strict ordinance against anyone publicly uttering the word "treasure" and even levied fines against it. The town would open up shop early in the morning and then around 10 a.m. the

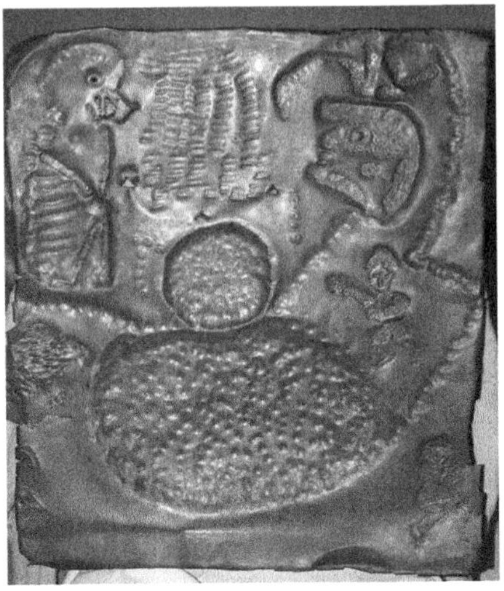
Aztec gold plate

stores would close, the doors were locked, and the streets became deserted as everyone headed up Johnson's Canyon and on to White Mountain.

Unfortunately, the treasure hunting came to an abrupt halt when it was discovered that Freddie's tunnel was full of sand. The Aztecs frequently built false tunnels to confuse thieves, and it appeared this was one of them. Freddie and his group of treasure hunters eventually became frustrated and gave up their search.[2]

The story of what happened to this mighty endeavor has made news in just about every treasure book. However, that doesn't stop folks from still looking. Even today, coins and other artifacts are found along the route between Kanab and White Mountain. Several people have been to Kanab to make films about Montezuma's treasure.

The metal detecting community is still finding bits of that past history as well—not far from the mouth of Johnson's Canyon, in the first little draw on the left, a stash of gold and silver was hidden by some robbers who forgot exactly where it was they hid their loot.

Somewhere in that little draw, a cache awaits some lucky treasure hunter.

NEPHITE GOLD ON THE OLD INDIAN CROSSING

In 1852, Chief Walker (Wah-kar-ar), then war chief of the Ute Tribe, was traveling back from a raid to the south with his booty and men. As they reached the Colorado River at the Old Indian Crossing it was getting late, so they had to find a place to camp as soon as they reached the other side of the river. As they began to set up camp, the mild weather they were enjoying suddenly turned to a driving rain that caused the men to seek shelter in the nearby rocks and crevices high about the river. By chance, Walker entered a small cave that afforded him and a few others comfort from the howling wind and driving rain. It was in this tiny cave that Chief Walker made a surprising discovery. There on the floor of the cave were hundreds of strange coins that he had never seen before. He and his men gathered many of the coins and stored them in their bundles, perhaps thinking they would make good trade items.

After many days, they arrived at Parowan where, for one reason or another, Walker singled out James H. Martineau and showed him a few of the strange coins he had found in the cave. Martineau took one of the coins and after rubbing off the grime and dirt was surprised by its luster. He knew right then and there that he was holding an ancient gold coin. According to one source, who has asked to remain anonymous, Walker told Martineau that the coins were only a part of what was found in the cave. There were other objects as well: some were of curious design while others were heavy and odd-shaped. Martineau, being a faithful member of The Church of Jesus Christ of Latter-day Saints, was certain he knew the origin of the coins and other artifacts in the cave. It was his belief that the coins and other items were the product of the lost Nephite nation that once inhabited the land. According to LDS doctrine, the Nephites came from Jerusalem about 600 BC.[3] Martineau believed the people settled first in the northeastern part of the country and then spanned out over the centuries.

Another source, Mad Bear of the Hopi Tribe, maintains that from a description of the country given to Martineau, he judged the cave to be in the area where the Colorado crosses the Arizona-Utah line.[4] But Chief Walker seemed to direct Martineau toward the Escalante River in present-day Cottonwood Gulch. However, Walker declined to take Martineau to the cave so its whereabouts are really anybody's guess at this point. Walker did allow Martineau to keep a coin to show Brigham Young, and that's what he did. Martineau sent the coin by mail to the LDS leader for his opinion. After months had passed, and he had heard no word from Brigham Young, Martineau had an inquiry through his bishop, who told him to forget that he had ever seen the coins, and to mention them to no one. He was

A sample of the coins found by Chief Walker

told that the day would come when all these things would be made known to him and to the Church.

KIVA ALONG THE ESCALANTE RIVER

In 1997, Nina Roberson, of Green River and who grew up in Escalante, was in an interview with me about a topic unrelated to hidden treasure when a story crept out about two young boys who liked to go a few miles out of town along the Escalante River to "jump cliffs" (rappel with ropes). It was in about 1950 when these two young men hiked down the Escalante Gorge along the west side of the river looking for a great place to rappel. They saw a strange-looking flat rock several dozen feet below them on a ledge. Anchoring their ropes to a large boulder, they ascended to the flat rock with ease. They quickly noticed that when they stepped on the flat rock their weight caused it to wobble, so they tied ropes to it and began dragging it aside to see what it was perched on.

The boys managed to dislodge the large boulder from its perch and saw a square hole going down into the earth. Upon closer inspection, they saw the flat rock had been chipped by mechanical means to make its shape. After inspecting the squared hole, it was obvious that it was man-made as well. There was a ladder at the entrance that allowed the two boys to enter into the kiva (an underground chamber made by the ancestral Pueblo). To their utter surprise, they discovered little stone boxes filled with gold and copper tablets with strange characters carved on them. They secured several tablets and one stone box and exited the kiva. They pulled the stone back over the hole and climbed up to the top of the ledge from whence they came.

Once back home, they contacted one of their fathers, who immediately told them they had disturbed an ancient site that they had no business in. Contacting the stake president, the boy's father was told to have them take him and other leaders to the site so the place could be forever hidden from view and so they could return the tablets to their rightful place in the kiva.

The five-mile hike took place a few weeks later, but when the boys thought they were at the spot where they had tied off their ropes to rappel to the large flat stone, nothing could be found that resembled it. They searched the reminder of that day and two days more without ever encountering the site. Some say the boys were lying and created the tablets themselves, but those who knew the boys believed them. According to

Mrs. Roberson, the tablets and the box were sent to Salt Lake City and the First Presidency of the Church for safekeeping.[5]

Was the cave Walker found full of old gold coins and other artifacts connected to the kiva the boys of Escalante discovered? Was it a Nephite cache site or perhaps evidence of another culture that visited the southwest?

NOTES

1. Rufus Wood Leigh, *Five Hundred Utah Place Names: Their Origin and Significance* (Deseret News Press: Salt Lake City, 1961), 45–46.
2. David Hatcher Childress, *Lost Cities of North & Central America* (Stelle, Illinois: Adventures Unlimited Press, 1992), 483. See also "Johnson Canyon and Montezuma's Gold," *WildernessUSA*, accessed April 10, 2017, wildernessusa.com/explore/articles/johnson-canyon-montezumas-tunnels/.
3. "Book of Mormon Time Line," *Ensign*, October 2011. See also "Kane County," Floyd Mann, *The Treasures of Utah,* forum, accessed April 10, 2017, thetreasuresofutah.yuku.com/topic/975/KANE-COUNTY#.WOw IolMrKis.
4. Zula Brinkerhoff, personal document in author's possession.
5. Nina Roberson, interview by Stephan Shaffer, 1997.

Millard County

"Millard County and its seat, Fillmore, were named for President Millard Fillmore. President Fillmore signed the Act creating the Territory of Utah, September 9, 1850, and appointed Brigham Young first Governor. Millard is one of Utah's largest counties, extending from the crest of the Pah Vant Plateau and its northern extensions on the east, to the Nevada line on the west; and its north-south dimension is nearly seventy miles."[1]

LOST MINE NEAR SEVIER LAKE

Millard County is not without its history of lost mines and hidden treasures. One such treasure story was penned by the late Delta City reporter Frank Beckwith, concerning the appearance of an old Mexican man who had discovered a rich ledge of gold in the desert.[2]

Pedro was a ranch hand working for a local rancher and had come north with an old Spanish map in hand looking for a lost mine his great-grandfather knew about. Pedro became fast friends with one of the other ranch hands. Once, while the two were hanging out together, the ranch hand asked Pedro to tell him about this lost mine. "Pedro got out a paper bag, smoothed it, and drew. He drew an oval, saying 'Laguna.' Another, saying 'Laguna, two and a half days west.' Then he drew peaks, 'Map say peaks, maybeso *[sic]* this mountain; señor, what you think?'"[3]

The ranch hand replied that the first oval could be Fish Lake and two and a half days' travel to the west would put him in Sevier Lake, with Notch Peak right where this map indicated a mountain.

His friend asked to keep the hand-drawn map, thinking he might find something, and Pedro let him take it. Later on, the ranch hand told Pedro that he had come upon a "funny-looking hole" and ridden on, thinking he had just discovered the lost mine.[4]

It took another map, however, to arouse any real interest in the hole.

Joe Nielson, a local cattle rancher, was given another map by Pedro and was determined to follow it to the mine.

Nielson's map was a far cry from the handmade one drawn on a paper bag. He, his son (Lafe), and his son-in-law (Alva Barnes), all got to see it. It was "an old map, frayed, beginning to part on the folds, a genuine map from Mexico—to the unbelieving [it was] merely the old stock-in-trade for shrewd swarthy sharpers to sell to 'Americanos'—'for a price, señor, maybe [some] riches for you.' But to believers, [the map was] an infallible certainty to previous workings of gold by the ancient Spaniards. [This map] too, showed two lakes, two and a half days travel apart; but it was marked with a hole, a tree, in dead line with a distant crag on the mountains."[5] It seemed the paper bag map and the Spanish map were dead look-alikes.

Joe Nielson claimed he found the "funny looking hole . . . on an alluvial fan, down a distance from the mouth of the big canyon. Joe promptly located it under the euphonious title, 'The Lost Josephine.'"[6] (There are several mines named Josephine in many mining districts of Utah.)

According to Nielson, his hole was "a nearly round oval, sixty by seventy feet and eighty-two feet deep. The sides were vertical: but it was in an unconsolidated, late alluvial deposit of the big fan, and soft. To avoid the danger of cave-ins, he dug a man shaft down the side of it . . . and then drifted a tunnel to the main hole."[7] From that point he fashioned a ladder that was about forty feet high and was said to still be in place in the 1950s. His search for the lost mine was fruitless; he finally gave up on his quest to become rich from Spanish treasures.

As for Joe Nielson's son-in-law, Alva Barnes, he discovered a partially filled-in hole about a mile farther up canyon on the edge of the canyon mouth. He called his hole, "The Lost Isabel," after the former queen of Spain.[8]

Alva struck pay dirt in his Isabel hole at a hundred and eighteen feet where he discovered cracks containing gold flakes. "He recovered several vials of the precious yellow stuff." However, as luck would have it, a flash flood came down the canyon at the height of his operation, filling the hole full of debris and washing away any hope of reopening the hole without a vast amount of capital. "Today, all that remains of his shack on the high land across the gulch is a battered up cook stove, a warped bed springs, and the usual flotsam of housekeeping."[9]

Another report of the mine near Sevier Lake came from a Dr. John E. Stains, a dentist in Delta, Utah. Pedro (possibly the same one as before) apparently went to him one morning with some gold he wished to sell to him. When Dr. Stains agreed to buy the gold upon inspection, Pedro poured out on a table the contents of his pouch.

Dr. Stains carefully tested the yellow metal and said, "'Malleable all right; works right. Then taking a larger pellet, he cut it in two with his knife. 'Same all through. No wash or plate on that.'"[10]

Then the dentist said he wanted to make one more test on the samples, so he applied the acid test that proved it was pure gold.

Pedro wanted five pesos for the pouch of gold. Dr. Stains knew it was underpriced and paid Pedro the five pesos immediately. Pedro said that he had about twenty-three pounds of it. This startled the dentist. Pedro went outside, then came back in with a bundle under his arm. He dumped its contents out on the table, whispering, "It all same. All gold." Dr. Stains was over cautious and worried that someone might kill Pedro for that much gold and told him so.[11]

Even though they tried to keep it secret, it still got out that Pedro had gold and people flew into the mountains, looking for Pedro's mine. Some think it was the Isabel hole while others believe it was located elsewhere. Pedro took his gold and headed for Los Angeles, never to be heard from again. But he told Dr. Stains where he got the gold before leaving. It's not known whether Dr. Stains ever went looking for the gold mine, but he never seemed to be without money, fine clothes, or the necessities of life.[12]

SILVER CACHE IN THE DRUM MOUNTAINS

Whether they are true or not, many stories seem to surface when one gets to talking about treasures and outlaw caches around the campfire. One such story was told to a group of prospectors camped out near Vernon Creek Reservoir in Tooele County. This is the tale:

> There was a shipment of silver bars coming from Nevada to Eureka, Utah, that consisted of two wagons. Each wagon was loaded with seventy-five twenty-pound silver bars and had an armed escort as well.
>
> However, it seems that a band of outlaws had been watching for this shipment of silver for some time and had picked an ambush site

along the way. Since the desert from the Nevada border to the House Mountains is nothing but open country, the outlaws decided to ambush the wagons at Marjum Pass in the House Mountain Range.

As the train of wagons and men approached the pass, the outlaws strung themselves out in the boulders and trees and waited. When the wagons had crested the pass, they attacked from both sides, killing the drivers of the wagons, two of the guards, and wounding two others.

Two of the outlaws hopped aboard the wagons and headed out toward the old town of Joy thirty miles to the northeast. They weren't in any big hurry because they thought that by the time their heist was discovered, they would have cached in the silver and escaped to parts unknown. Little did they know, one of the wounded men was able to use two horses to make it back to the telegraph station at Baker, Nevada, where he instructed a message to be sent to Ely, Nevada, and Delta, Utah.

The sheriff of Delta wasted no time in mustering a well-armed posse and headed toward Joy as fast as he could go. Another posse was formed in Baker and they too headed out following the trail of the wagons.

When the posse from Baker came into Whirlwind Valley, between the House Range and the Little Drum Mountains, they could see the dust of the stolen wagons far off in the distance. Using a telescope, they were assured that the wagons they saw were indeed in the hands of the outlaws.

Meanwhile, the Delta posse had made it within a few miles of Joy when they saw a group of men on horseback coming their way. As soon as the group saw the posse, they turned their mounts around and headed back the way they came, riding as fast as they could go, with the posse in hot pursuit.

As they came into the little hamlet of Joy, they were met by the Baker posse and a shootout began. All the outlaws were killed except two who were seriously wounded. The wagons were discovered but without the silver bars. A search of the area, the road coming into Joy, and every wash and ditch was gone over a dozen times but still no silver. The posses questioned the two wounded men but they would not give up the information. The two remaining outlaws were sent to prison in Nevada. One outlaw died while inside and the second was released after serving ten years.

It was a month or so after the second outlaw's release that he was spotted in one of the abandoned rock houses that still stood at Joy. The Millard County Sheriff would ride out to see what he was up to a few

> times a month and each time, the old outlaw seemed to have a few more items for his dwelling and his personal needs.
>
> The sheriff knew that he was getting silver from somewhere, but where? He knew about the hidden bars and seeing that the old outlaw wasn't in the best of shape he also knew the bars had to be somewhere within easy reach. But the cache was never found. The last time the sheriff paid the old outlaw a visit he found the man's dead body in his rock house. All around the house were holes where someone had dug for the hidden silver bars. The sheriff buried the remains of the old man near the rock house. Later, he found a saddlebag near the house with four squared-off pieces of pure silver.[13]

Perhaps the rest of the silver cache can be found in one of the many crevices that opened up through the "drumming" of the Drum Mountains. Or perhaps the cache is completely buried by the same mountains. It's still worth hunting for.

LOST TREASURES OF THE CANYON MOUNTAINS

The Canyon Mountains are situated east of Oak City and have always been an exciting place to prospect, recreate, and more. Many stories of lost mines and treasures persist in these mountains, and folks still hunt for those lost mines and treasures to this day. Stories have surfaced of a few significant finds such as Spanish coins, bits, spurs, and carved trees and rocks. Outlaw stories tell of hidden caches along the foothills at Whiskey Creek.

The Old Spanish Trail cuts across the very bottom of the Canyon Mountains. Two Dominguez-Escalante monuments have been erected on both the west and east side at the bottom and between the Church Mountains. The old trail was used by Indians, Spaniards, outlaws, and pioneers. It stands to reason why stories of old mines exist in that area.

SPANISH MINE ON WILLIAMS PEAK

A Ute Indian by the name of Lawrence Walker was poking around the north of the Old Spanish Trail, hunting for a mine his grandfather had told him about. His grandfather said it was a good gold mine that the Spaniards would take gold out of when a certain group of them were in the area. He told of two streams that began from the same source: Whiskey Creek and Eight Mile Creek.

Walker hiked the area between these two creeks off and on for years. Although he never found the mine, he did find many markers and monuments. He told me that he knew he was close because of all the markers he had found. He figured the mine had been sealed shut either by the Spaniards or by Indians. He drew a map, but it is hardly to scale or even close to what the geography of the area is; this is typical of most maps drawn from memory.

The most likely place for the mine to be would be on Williams Peak. The Whiskey Creek headwater is on the west side of the peak while the Eight Mile Creek headwater is on the east side. Walker marked two more sites where he was told the mine could be located.

Somewhere near the two creeks, a lost Spanish mine lies waiting for some lucky explorer.

GOLD IN SEVIER DESERT

Another campfire story tells about a prospector in 1869 by the name of Frank Lane, who "found large quantities of gold in potholes [natural pits created by water currents] which he discovered while searching for water on top of a long six-foot-high ridge in the Sevier Desert between the Confusion Range to the west and the House Range to the [east]. Lane removed $100,000 from a single pothole and returned to his home in the East. Years later, he returned penniless and spent years searching for the site without success."[14]

TABERNACLE HILL'S VANISHING LEDGE

Years later, around 1880, a cowboy by the name of Hill "became lost in a sandstorm somewhere near the north end of the Low Cricket Hills and took shelter in a cave-like opening in a rocky ledge which stood about ten feet above the desert floor. Inside, he found sacks of gold nuggets and dust and a store of smelted bars with cross markings. When the storm ended, he stuffed his pockets with as much gold as he could carry, and after a day of walking . . . southeast, saw the outline of Black Rock Volcano [now known as Tabernacle Hill]." From there he made it to Fillmore, near dead from lack of food and water. After he rested, he headed back to the treasure cave, but somehow the mysterious ledge had simply vanished! He never did find it.[15]

HERSCHEL'S THREE CROSS TREASURE

"A man named Herschel discovered a cave, whose floor was covered with gold that resembled grains of wheat." The cave was located about twenty-five miles west of Fillmore and within a few miles of the eastern edge of the Black Rock Desert. Herschel "carved three different crosses on the rocks around the cave, but after being attacked by outlaws, he fled the area and was later unable to relocate the site upon his return."[16] He was certain the outlaws knew nothing about the cave.

"THERE'S GOLD IN CORN CREEK!"

"There's gold in Corn Creek!" was the yell. During the 1850s, settlers of Fillmore and Hatton frequently reported pack trains of ore being brought out of Corn Creek Canyon by Spanish miners. Some of the settlers asked the Spaniards what they were up to. Those who wished to tell them said that they had been working in mines all over the area—mines that had been worked by their families for generations. They were taking gold ore to a mill further south for processing.

During the two Indian Wars (Walker and Black Hawk), many of the ore trains were attacked by Indians and many miners lost their lives. The mines were lost to history until early in 1900, when a man from New Mexico named Robert Keen came north with one of the survivors of an Indian massacre at a mine in Corn Creek.

The survivor's name was Joe Gutierrez, and he was only eight years old when he witnessed the death of his father and other miners. Gutierrez was one of only seven who escaped the slaughter.

Keen talked Gutierrez into returning to Utah to find the old mine, since it was said to still retain rich ore deposits and Keen wanted a piece of it.

Finding the old mine was much harder than they first thought. Things changed over time; brush and trees grew in new places, and where there was once brush and trees, now was bare ground. Gutierrez had to find all the old landmarks from memory. But after two years of searching, they found his father's mine at the head of Corn Creek, east of present-day Kanosh.

Here they discovered several sunken depressions in the ground, which proved to be the old Spanish shafts, cleverly covered by timbers, grass, and sagebrush. Some of the old tunnels had been concealed by large boulders in order to make it look natural. The men uncovered one shaft and constructed a windlass hoist over the top of it, so they could lower themselves into the bowels of the mine.

When the men descended into the shaft, they found the mine had different levels. Each landing was made from large flat stones set into the sides of the shaft in fitted grooves. Steps had been notched into the shaft walls, worn down by Indian slaves who had climbed them with heavy packs of gold ore.

Keen had some of the ore he collected from the shaft that assayed at $22,000 per ton. However, the old tunnels and shafts were so dangerous to work in, he decided to drill a new tunnel from lower on the mountain to connect with the bottom of the main shaft.

While digging the new tunnel, a main ore body was struck. The *Salt Lake Mining Review* reported it is as being, "A large body of telluride ore, with average values of $263 to the ton, although some of the ore assays more than $1,400, this begins to look encouraging!"

Keen and Gutierrez, along with several prominent mining men, organized the Old Spanish Mining Company. They worked the mine until high water levels in the shaft forced them to abandon it.[17]

Some folks from Kanosh still claim that there are caches left on the mountain, as well as scattered gold ore left over from the many Indian attacks on the Spanish caravans coming out of Corn Creek.

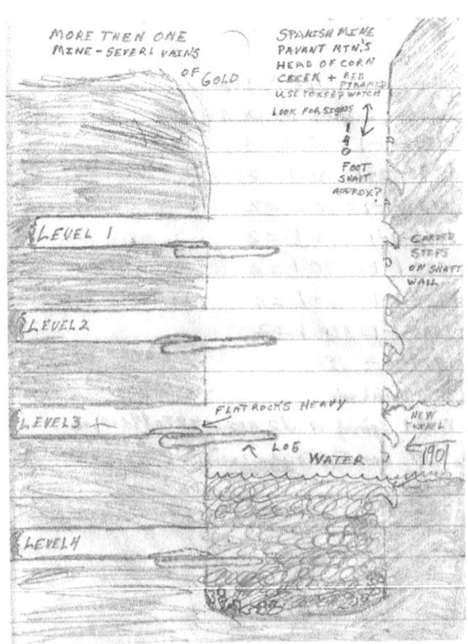

Hand sketch of the shaft

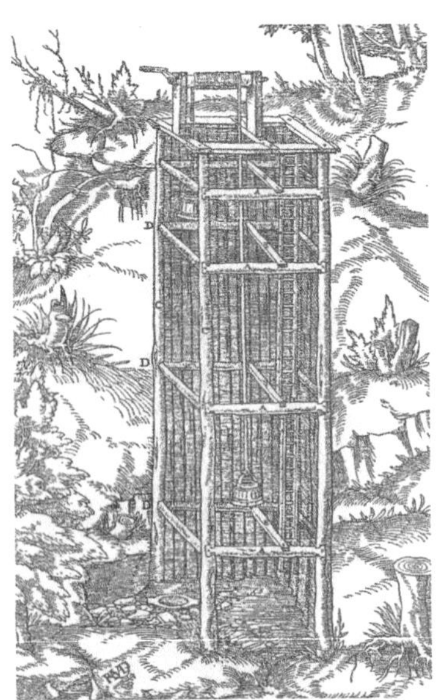

Ancient shaft design

NOTES

1. Rufus Wood Leigh, *Five Hundred Utah Place Names: Their Origin and Significance* (Deseret News Press: Salt Lake City, 1961), 58.
2. Reported by Frank Beckwith. See also Frank Beckwith, "Pedro's Lost Gold Mine . . . ," *Desert Magazine* (April 1951), archive.org/stream/Desert-Magazine-1951-04#page/n21/mode/2up/search/beckwith.
3. Frank Beckwith, "Lost Spanish Mine," in *Rejected Manuscripts*, collections.lib.utah.edu/details?id=125394#t_125394.
4. Ibid.
5. Ibid.
6. Ibid.
7. Ibid.
8. Ibid.
9. Ibid.
10. Ibid.
11. Ibid.
12. Ibid.
13. Daryl Hendricks, story told to author.
14. "Millard County," Floyd Mann, *The Treasures of Utah*, forum, accessed April, 10, 2017, thetreasuresofutah.yuku.com/topic/1000/MILLARD-COUNTY#.WOwCbFMrKis.
15. "Millard County," Floyd Mann, *The Treasures of Utah*, forum, accessed April, 10, 2017, thetreasuresofutah.yuku.com/topic/1004#.WOwEDFMrKis.
16. "Millard County," Floyd Mann, *The Treasures of Utah*, forum, accessed April, 10, 2017, thetreasuresofutah.yuku.com/topic/1001/MILLARD-COUNTY#.WOwMWFMrKis.
17. Story told to author by the great-great grandson of Bill Burns, who worked the mine.

Morgan County

"Morgan County is in the heart of the Wasatch Mountains; it embraces the Weber River Valley west of Devil's Slide to the mouth of Weber [Canyon]; and southerly, the historic East [Canyon]. Morgan County was organized in 1862, and named for Jedediah *Morgan* Grant, Mormon pioneer, counselor to Brigham Young, and father of Heber J. Grant, president of [The Church of Jesus Christ of Latter-day Saints]. The town of Morgan in the Weber Valley is the county seat."[1]

THE THOUSAND MILE TREE

If there is anything Morgan County is famous for, it is the "Thousand Mile Tree." This pine tree, located in Weber Canyon along the Union Pacific Railroad route, marks exactly one thousand miles of rail from Omaha, Nebraska, to Council Bluffs, Iowa—"The eastern terminus of the First Transcontinental Railroad."[2]

Utah rail historian Don Strack described how the Thousand Mile Tree reached its fame in his article, "Eastbound To Wahsatch—Union Pacific's Route Through Weber and Echo Canyons." According to Strack, within a week of 15 January 1869, the "tracks reached the site of a large tree, 90 feet tall, that happened to be exactly 1,000 miles from Omaha, and soon a sign was hung from the tree clearly stating that fact."[3]

Strack continued:

> The tree was in middle of a gorge between Henefer in Upper Weber Valley and Devil's Slide, a unique geological formation of twin limestone ridges running vertically from the canyon floor. Along with

> the Thousand Mile Tree, Devil's Slide immediately became a sight to be seen by all passing trains. The gorge just east of Devil's Slide was named Wilhemina Pass and was the subject of several views by Union Pacific's official photographer A. J. Russell for his stereographic tour of the new line. Although the gorge was changed significantly to accommodate today's Interstate 84, most early trains stopped to allow passengers to appreciate the landmark, and several excursion trains from Ogden were arranged to see Wilhemina Pass, the Thousand Mile Tree, and Devil's Slide.[4]

The Thousand Mile Tree died thirty-one years later, and it was removed in September 1900. The rail line was later modified, reducing the exact mileage to the site by approximately forty miles. But in 1982, Union Pacific planted a new tree in commemoration of the original. The new tree is now over thirty feet tall and stands behind a special enclosure.[5]

THE SPANISH MYSTERY OF MORGAN COUNTY

Several exciting discoveries have been made in Morgan County, particularly the discovery of an ancient Spanish cache near Porterville. While deer hunting on private property, some local residents happened upon some artifacts in a lonely canyon south of town. It is unknown how the items came to be in the canyon. Two mines were eventually discovered, as well as a strange-looking lead map that seemed to "show the way" to other hidden mines and caches.

Many folks have tried in vain to access the property but it is not open to the public. Only a select few have ever wandered the hills south of Porterville undisturbed.

SPANISH CANNONS NEAR SADDLE ROCK MOUNTAIN

During a search for other mines in the vicinity, I met a man by the name of Mark Mason from Ogden. Mason told me about a man he met from Hill Air Force Base who claimed he had discovered two Spanish cannons southwest of East Canyon Reservoir near Saddle Rock Mountain.

Hunting deer, he climbed to a higher vantage point where he thought he would be able to view the country better. Suddenly, while moving through a stand of quaking aspens, he came upon a cannon barrel sticking up out of the ground. Soon after, he discovered another one not far away.

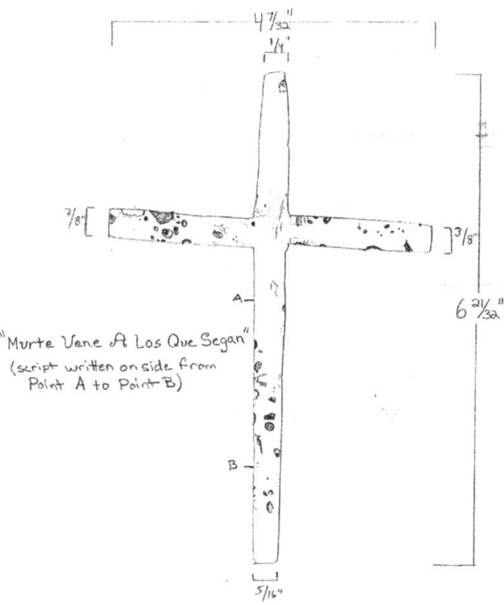

Morgan County's Spanish mystery map #!

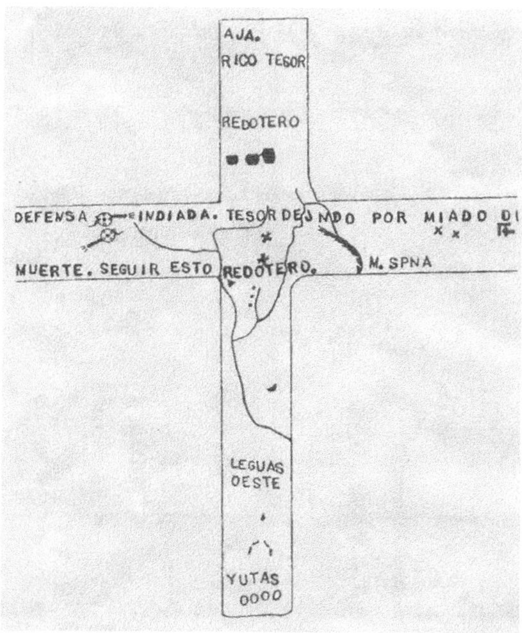

Morgan County's Spanish mystery map #2

Very excited upon this discovery, he immediately decided to cover them better in case another hunter ventured onto the spot. He cut branches from the surrounding stands of aspens and covered the two cannons the best he could. Later, he confided in Mason their location.

I, along with a former partner, made a few attempts to locate the hidden cannons with no success. Are the cannons really there? Were they removed? Perhaps someone will make the attempt to search the mountains again and find these rare gems.

NOTES

1. Rufus Wood Leigh, *Five Hundred Utah Place Names: Their Origin and Significance* (Deseret News Press: Salt Lake City, 1961), 61; italics in original.
2. *Wikipedia*, s.v. "Thousand Mile Tree," last modified March 27, 2017, en.wikipedia.org/wiki/Thousand_Mile_Tree.
3. Don Strack, "Eastbound to Wahsatch: Union Pacific's Route through Weber and Echo Canyons," *The Streamliner* 17, no. 3 (Summer 2003), utahrails.net/articles/weber-echo.php.
4. Ibid.
5. *Wikipedia*, s.v. "Thousand Mile Tree," last modified March 27, 2017, en.wikipedia.org/wiki/Thousand_Mile_Tree.

Piute County

"Piute County was split off from Beaver County in 1865. Its western boundary approximates the crest of the Tushar Mountains. . . . Most of the [county's] population is concentrated in the Sevier River Valley."[1]

THE ANNIE LAURIE GOLD COMPANY

Mining played a major role in the very survival of Piute County. After the arrival of Mormon settlers, some began prospecting the hills just west of Marysvale in the Tushar Mountains.

Gold was discovered as early as 1888; silver and other ores were discovered a little earlier.

Early mining at Kimberly

Newton Hill located the famous Annie Laurie mines . . . in 1891, and Willard Snyder developed the Bald Mountain Mine. Snyder platted out a Mill Canyon townsite, which he named *Snyder City*. . . . The real growth began in 1899 when Sharon, Pennsylvania investor Peter L. Kimberly bought the Annie Laurie and other area mines. Kimberly incorporated his holdings as the Annie Laurie Consolidated Gold Mining Company, which established a gold cyanidation mill there.

The town, renamed *Kimberly*, began to boom. . . .

Kimberly quickly became the leading gold camp in the state, with two hotels, two stores, three saloons, and two newspapers. In 1900 the county formed the Gold Mountain School District, and a log schoolhouse was built.

The boom period of 1901–1908 is considered to be the town's heyday; the Annie Laurie Company absorbed several other mines and paid out nearly $500,000 in dividends. . . . By 1902 the Annie Laurie employed 300 miners, and Kimberly's population reached 500. . . .[2]

HOME OF BUTCH CASSIDY

One of the most celebrated outlaws in Utah's history grew up in Piute County. George LeRoy Parker, aka Butch Cassidy, and his family lived just outside of Circleville in northwestern Piute County.[3]

Many years ago, while doing some research in Circleville, I stopped in to visit with Butch Cassidy's ninety-five-year-old sister, Lula Parker Betenson.

The boyhood home of Butch Cassidy

She had recently helped Robert Redford in his story about Cassidy and the Sundance Kid.[4]

Mrs. Betenson related to me her take on what really happened to her brother so many years before. But first she told me about the last time she saw her brother. It was at the death of her mother when Butch and some members of the Wild Bunch came to town to attend the funeral.

The town marshal told Butch that he was welcome to stay for the funeral services, but as soon as it was over, he and the Wild Bunch had twenty minutes to get out of town before he would come after them with a large posse of men, horses, and guns. True to his word, just as soon as the time was up, the marshal and his posse headed out of town after the gang. Mrs. Betenson didn't know if any of Butch's men were caught, but she did know Butch made a clean getaway.

She told me that Butch never went to South America, but abandoned his life of crime and went to work for the railroad in Washington State where he worked until he retired. Later, she received a letter from a man from Pahrump, Nevada, who had become a close confidante of Butch. The letter said that Butch had moved there after he retired and became sick. Knowing he was near death, Butch asked his friend to write to her and let her know where he was.

Butch died in Pahrump but was somehow brought back to Utah, and is now buried near his old homestead in an unmarked grave.

Mrs. Betenson told me that many people thought she knew where some of the Wild Bunch's cache sites were, but she didn't know anything about them. She assumed they had cache sites all over the country but didn't have a clue as to where.

THE CANYON OF GOLD

> Carved into the eastern slope of the Tushar Mountains of south-central Utah, rise[s] the walls of Bullion Canyon. Surrounding mountain peaks soar more than 12,000 feet above sea level. Pine Creek descends through the gorge, losing more than 2,000 feet of elevation until it merges with the Sevier River about four miles west of the canyon. At the confluence sits the town of Marysville at 6,000 feet. Vegetation ranges from stands of aspen, ponderosa pine, and Douglas fir on the north-facing canyon slopes to sage brush at the foot of the mountains.
>
> Indians of the Fremont culture inhabited the region about a thousand years ago, leaving little but petroglyphs and pottery shards.

Ute and Paiute roamed the area by the eighteenth century. The first Europeans to discover the Tushars were Spaniards, seeking gold. An arrastra stone, used in crude milling of gold ore, still lies next to Pine Creek, providing mute testimony of early mining. Jedediah Smith, the first American to see the Tushar Mountains, probably saw only their western slope. Not until the 1850s did Mormons pass through the area.

The history of Bullion Canyon is a history of mining. From the first rush in 1868 until the mines died out a hundred years later, men sought precious metals in the gorge. While the motive that drew men to live in the canyon remained the same, the character of the settlements over time changed radically. Initially wild and unrestricted, canyon settlements grew increasingly civilized as families formed.

In the summer of 1856, George A. Smith, the second counselor to President Brigham Young, explored south-central Utah, following the Sevier River until he encountered a winding canyon just south of the confluence of Clear Creek and the Sevier. Probably to avoid that twisting obstacle, Smith ascended what is now known as Monroe Divide and became the first recorded member of the Mormon community to look down on the valley from the eastern foot of the Tushars. His glance took in the ribbon of green cottonwood foliage that spilled out of the Tushars and merged into the larger lush strand that marked the course of the Sevier. Smith and his party camped at the confluence of the mountain stream and the Sevier. During a stay of several days, the group engaged in revelry that culminated in a "stag dance" [a practice where, if no women were available, men danced with other men]. Smith named the locale "Merry Vale" after the festive mood of his men. In time, the name was changed to Marysville.

Among Smith's party was a tough frontiersman, Sylvester Hewitt, who had been present at the initial discovery of gold in the Sacramento Valley in 1848. Hewitt was a member of the Mormon Battalion and a competent, daring prospector. While at "Merry Vale" Hewitt panned Pine Creek and found traces of gold dust. Word did not spread as rapidly as it might have elsewhere. . . . Brigham Young discouraged prospecting and mining, because he believed his people should first establish themselves in agriculture and other stable enterprise. Consequently, Smith tried to prevent the news from being spread by demanding silence from the party.

Historically, the discovery of gold has been a notoriously difficult secret to keep. In 1868, Lieutenant Jacob Hess and Ebenezer Hanks, probably having heard of Hewitt's discovery, rode into the Marysville country in search of gold. They found color in Pine Creek and tried sluicing to extract the metal from creek gravels, but the concentrations were too small to be profitable. The men were earning only a dollar a day.[5]

Remnants of that old mine and town site can still be accessed today; it is a fun place to metal detect.

NOTES

1. "Piute County, Utah," *Utah's Online Library*, accessed April 11, 2017, onlinelibrary.utah.gov/research/utah_counties/piute.html.
2. *Wikipedia*, s.v. "Kimberly, Utah," last modified March 22, 2017, en.wikipedia.org/wiki/Kimberly,_Utah; italics in original.
3. *Wikipedia*, s.v. "Butch Cassidy," last modified April 2, 2017, en.wikipedia.org/wiki/Butch_Cassidy.
4. Lula Parker Betenson. interview by Stephan Shaffer, 1972.
5. "A History of Bullion Canyon," Daniel Glass, *Marysvaleutah.org*, accessed April 11, 2017, marysvaleutah.org/informationaboutmarysvale/history/59-a-history-of-bullion-canyon-marysvale-utah-piute-county.html.

Rich County

"Rich County embraces the northeast corner of Utah, being a long, narrow area extending northward from Echo [Canyon]. The southern half of Bear Lake is in Rich [County]; the northern half, in Idaho. Rich County was organized in 1864 and named for Charles C. Rich, an early Mormon apostle, who was prominent in the settlement of the Bear Lake region."[1]

CACHE OF THE ROCKY MOUNTAIN FUR COMPANY

> The Rocky Mountain Fur Company headed by Milton G. Sublette, Dave E. Jackson, and Jedediah S. Smith conducted a fur trading rendezvous in the vicinity of Laketown on July 1827, taking 130 bales of beaver furs for shipment to St. Louis in March with 60 men and merchandise arriving via South Pass in late June. The trading was concluded and all parties dispersed in mid July 1827, following the return of Smith from a perilous journey to California. Traders [visited again] in 1826. In the late 1860s, Meadowville, Round Valley, and Laketown were being established as "Mormon" communities in spite of troubles with the Indians over . . . their "hunting grounds." It seems that the first white settlers in the valley had made a treaty with the Indians, which gave to the whites the north end and the Indians the south end of Bear Lake Valley. Large bands of Indians frequently gathered in the vicinity of Laketown. In 1870, a gathering of Indians, (estimated at 3,000), camped on the south shore of Bear Lake causing settlers a great deal of concern and trouble; however, after a meeting of the settlers and chiefs, among them Chief Washakie, an agreement was effected and the Indians moved to Wind River, Wyoming. Today, this same area is

a thriving ranch community with an LDS church, elementary school, stores, service stations and cafes. On the shore of the lake are many permanent and recreational homes. The Cache National Forest, which borders the Laketown community on the west and south, abounds with deer and elk. To the north and east of the town, the Great Salt Lake Council of Boy Scouts has established an aquatic camp on lakeshore acreage.

The trappers and traders of the early 1800s were familiar with the Bear Lake country. One, Donald MacKenzie (1819), a red haired Scotsman, is credited with naming Bear Lake and Bear River, so called because of the numerous black bears in the area. He explored the country and traded with Indians. At MacKenzie's instigation, over 10,000 Indians camped on both sides of a seven-mile stretch of Bear River at the north end of Bear Lake. The gathering (including the Bancocks, the War-are-ree-kas, and the Shoshonis, not all of which were on friendly terms themselves) was the largest ever known to have assembled in the Rockies. The giant Indian Chief Pee-eye-em and his brother Ama-qui-em, also a huge man, were in authority over the entire group. Mackenzie's purpose in arranging for this [powwow] was not only to trade for furs, but also to persuade the Indians to be friendlier to the white people.

The first covered wagons came into the Rocky Mountains in 1830, during which year they made their way as far west as Fort Washakie in Wyoming. Continued efforts were made to find passable wagon trails through mountains to the Pacific Coast, which goal was finally reached in 1840. At that time, the entire northwest mountain area was known as the Oregon Country and western travel was either to the Oregon or California regions. While early maps give the probable location of the first Oregon Trail somewhat to the north, well marked wagon ruts and the stories of Indians and early settlers of this region indicate that the first wagon to "Migration Oregon" followed the southerly and westerly shores of Bear Lake, leaving this valley through a canyon to the northwest.

Additional color is given to this belief by the fact that this area was the site of an important trapper rendezvous as early as 1827, and well-marked trails were followed for many years in and out of the valley. The first permanent settlement in the Valley was Paris, Idaho, settled in the fall of 1863 by "Mormon" pioneers led by Charles C. Rich. Most of the other towns were settled the following spring.[2]

THE BEAR LAKE MONSTER

> Rumors of the Bear Lake monster were reported to the first white settlers of the lake by the Native Americans of the region. They reported that many times in the past it had captured and carried away braves who were swimming. They said the monster was of the serpent kind, but had legs about 18 inches long and sometimes crawled out of the water.
>
> From the first settlement of the valley by the Mormons in the early 1860s, various persons reported seeing a huge animal in the lake.[3]

The monster has been seen by many and has been described as having a mouth "large enough to swallow a man" and "huge ears." One account describes not one monster living in Bear Lake, but "four large ones and six smaller ones."[4]

Although the monster has been "discredited by a group of scientists . . . , the story of the Bear Lake monster . . . [has] been widely enjoyed throughout Utah."[5]

NOTES

1. Rufus Wood Leigh, *Five Hundred Utah Place Names: Their Origin and Significance* (Deseret News Press: Salt Lake City, 1961), 82.
2. "Rich County, Utah," *Rich County, Utah*, accessed April 11, 2017, www.richcountyut.org/documents/history/RICH%20COUNTY.pdf.
3. "The Bear Lake Monster," *Laketown Lodge*, accessed April 11, 2017, www.laketownlodge.com/pdf_new/Misc.pdf.
4. Ibid.
5. Ibid.

San Juan County

"San Juan County was created in 1880; it is the largest county in Utah—an immense domain having an area of 7,761 square miles in the southeast corner. Its northern boundary is common with Grand County; its western bounds are the Green and Colorado Rivers; on the east is Colorado; and on the south, Arizona. The topography of San Juan is varied, and in many areas weird and inaccessible. Running from east to west in its southern section is the San Juan River from which the county was named."[1]

The history of the San Juan River is important to this narrative because it is an essential part of the over-all history of San Juan County. The full Spanish form of the name [Río San Juan Bautista], translates to "Saint John the Baptist River."[2]

The San Juan River begins its journey near "Farmington, New Mexico, and courses northwesterly into Utah a few miles north of Four Corners, [then] generally westerly across San Juan County," ending at the Colorado River in the San Juan Mountains. "In its lower courses the San Juan has carved a deep box [canyon] with many entrenched meanders, and enters the gorge of the Colorado in Glen [Canyon]."[3]

San Juan is a name common in New Mexico, Colorado, and Utah.

There are two explanations of the origin of the name San Juan River: (1) In 1598 the Spanish, under Don Juan de Oñate, undertook the conquest and settlement of *Nuevo México*. Oñate possessed the Tewa pueblo *Caypa* on the east side back of Rio Grande del Norte north of Santa Fé as headquarters of the provisional Spanish government; he renamed the Indian pueblo San Juan, to honor himself, by which name it has since been known. . . . The name was supposed to have been applied to the mountains to the northwest and the river flowing from them. (2) According to J. J. Hill, it is possible Río San Juan was so named by *Friars Dominguez y Escalante* in honor of Don Juan María de Rivera who, in 1765, explored northwest from Santa Fé to what is now known as the Gunnison River.[4]

OUTLAW HIDEOUTS IN LA SAL

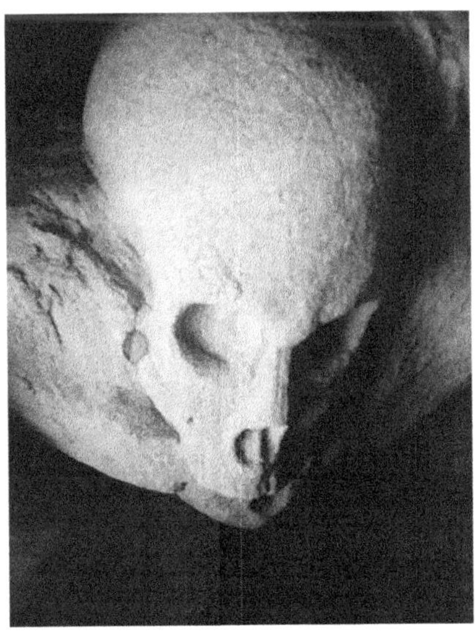

A skull protects a hidden cave in the La Sal

La Sal was a hideout for many outlaws in the [1880s] and 90s. Butch Cassidy, Matt Warner, Monte Butler, Kid Jackson, Al Ackers, Kid Parker, Bert Madden, Tom Roach, and the McCarty brothers were all known in this area. They were sometimes cowboys, ranchers and outlaws. These and others were known as the Wild Bunch and made their home in La Sal and Robbers Roost near Green River. Tom McCarty was married to Matt Warner's sister. These La Sal brothers helped rob the Telluride, Colorado bank in 1893. They were also cattle rustlers. They were connected to the Younger gang. Matt Warner, Tom McCarty, and Butch Cassidy joined forces and became known as *The Invincible Three*, and were notorious throughout the country. Tom was cool, clever, and bold. Matt was brave, but reckless, with an uncontrollable temper. Butch was cool and calculating. They made a formidable trio.[5]

SPANISH GOLD IN RECAPTURE CREEK

Many tales of lost outlaw loot, Spanish mines, and hidden treasure have been told around kitchen tables, campfires, and in bars and café's for many years. Just how true these stories are is for some young adventurer to find out.

There are stories that tell of gold in Recapture Creek. In the late 1880s, Andy Laney was watering his horse at Recapture Creek when he accidently found an eight-inch bar of gold. At a later date, he returned to the site and reportedly found several more bars.[6]

In 1964, a treasure hunter using a metal detector found a ten-inch-long gold bar on the same creek. And other accounts have been verified.[7]

Hunting for gold

$55 MILLION NEAR MOONLIGHT CREEK

There are other stories of buried treasure, like a huge cache of Spanish gold reported to be worth a whopping $55 million dollars buried somewhere in the area of Moonlight Creek.[8]

GOLD BETWEEN BLUFF AND MEXICAN HAT

Another story states that in about 1880, Abel Herringer worked a farm along the Comb Ridge in the area between Bluff and Mexican Hat. He discovered a rich ledge of gold in an isolated area of his property in a side canyon of the San Juan River. He secretly worked the site. By the time the ledge was worked out, he had accumulated $100,000 in gold that he put into leather sacks and buried at different locations on his farm. On his deathbed, he confided the story to a close friend who made extensive searches later. The friend never found any of Herringer's sacks of gold. However, one sack of raw gold was discovered by a man plowing what used to be the Herringer farm.[9]

GOLD DEPOSIT IN THE SAN JUAN

Prospector Jim Douglas found a deposit of fine gold on a sand bar in the San Juan River about eight miles northeast of Mexican Hat during a drought in 1909. He was flooded out during the spring runoff and waited twenty years for the river to reach the same low point so he could rework the same location. It never came, and Douglas committed suicide in 1929, taking the secret with him.[10]

THE "HOLE-IN-THE-ROCK"

One cannot write historical events of San Juan County without mentioning of one of the greatest feats known to take place during 1879. On April 13th of that year, an exploration party left Paragonah, headed for a little-known area east of the mighty Colorado River and near the San Juan River. The party consisted of twenty-six men, two women, eight children, eight wagons, eighty horses, and sixty-six head of loose cattle.

> The families of Harrison H. Harriman and James I. Davis, with four children each, were prepared to settle and start a colony wherever a suitable site could be found. Though they were leaving home and friends to pioneer a completely unknown land, they were anxious and eager to go.
>
> Traveling south, via Kanab, the explorers reached Lee's Ferry and crossed the Colorado River on April 30. Eastward over Lee's Backbone and across the Bitter Springs Desert, sometimes finding little or no water for up to fifty miles at a stretch. One-third of their cattle died of thirst before the company reached the Moki Indian village of Moenkopi, Arizona, where welcome rest and water awaited them.[11]

One of the explorers, Silas Smith, spoke with many of the Indians. He asked them how hard it would be to drive wagons and livestock two hundred miles through Navajo country to the San Juan. Those he had asked said it would be almost impossible to make such a trek with wagons, since water was so scarce and the country very inhospitable. Nevertheless, Smith was determined, so after only a week of rest, the company was on its way. Some days were so hot that the

Hole-in-the-Rock

sweating horses and oxen could barely pull their wagonloads. The Navajo demanded a heavy toll for the pioneers to pass through their land, but the company had to comply since they didn't want or need any trouble from the Navajo.

After eighteen hard-fought days, the company "arrived at the San Juan River near McElmo Wash. It had been sixty-one days since the expedition had left Paragonah."[12] The trip was a little over four hundred miles of mean, hostile country.

The exploration company made it back to Paragonah after having been gone for five months, and Smith gave a report to Erastus Snow. "Smith said that the southern route was impossible for a large group. There was not enough water or forage, and the Indians were hostile. Smith estimated it was only about two hundred miles on a straight line from Parowan to Montezuma and reasoned that there must be a faster, more direct route."[13] This was the beginning of the great trek to and through the Hole-in-the-Rock.

For more information about the Hole-in-the-Rock pioneers, check out *Incredible Passage: Through the Hole-in-the-Rock*, written by Lee Reay in 1980.

A vehicle trip along the old Hole-in-the-Rock trail is well worth the time and effort. There are many places to explore and reflect on the brave men and women who conquered this barren and unforgiving land. Prospectors and adventurers hunt for lost gold and silver on the old trail. Many items had to be left behind by the pioneers.

NOTES

1. Rufus Wood Leigh, *Five Hundred Utah Place Names: Their Origin and Significance* (Deseret News Press: Salt Lake City, 1961), 85–86.
2. Ibid., 86.
3. Ibid.
4. Ibid.; italics in original.
5. "History of La Sal, Utah Area as Given By Norma P. Blankenagel," *Moabmuseum.org*, recorded by Lula Delong, accessed April 11, 2017, www.moabmuseum.org/wp-content/uploads/2016/05/02-Primary-Subject-Personal-Oral-History.pdf; italics in original.
6. "Montezuma's Treasure-in Kannab—Re: Recapture Creek," Floyd Mann, *The Treasures of Utah*, forum, accessed April 11, 2017, thetreasuresofutah.yuku.com/topic/438/Montezumas-Treasurein-Kannab?page=2#.WO0La1MrKis.
7. Ibid.
8. "San Juan County," Floyd Mann, *The Treasures of Utah*, forum, accessed April 10, 2017, thetreasuresofutah.yuku.com/topic/973/SAN-JUAN-COUNTY#.WOwImlMrKis.
9. George A. Thompson, *Some Dreams Die: Utah's Ghost Towns and Lost Treasures* (Salt Lake City: 1982), 82–83.
10. John W. Van Cott, "Douglas Mesa," in *Utah Place Names: A Comprehensive Guide to the Origins of Geographic Names* (Salt Lake City: University of Utah Press, 1990), 114.
11. Lee Reay, *Incredible Passage Through the Hole-in-the-Rock* (Salt Lake City: Publishers Press, 1980), 13.
12. Ibid., 15.
13. Ibid., 19.

Sanpete County

"Sanpete was organized as one of the original eight [counties] when the Utah Territory was created by [the] Congressional Act in 1850. The Territorial Legislature corrupted the aboriginal name to this form, which may readily be mistaken for a corruption of 'Saint Peter.' Sanpete County includes San Pitch Valley and its watersheds."[1]

"San Pitch Valley is one of Utah's rich agricultural and thickly populated valleys. [Its name] derived from the name of the Ute division whose homelands embraced this area, and who were known by several variant names: *Sampitches*, *Sampichya*, *Sampiches*, and *Sanpuchi*. 'They wintered in the Sevier River Valley where there was less snow and where they could take deer, rabbits, ducks, and geese and trap beaver and mink."[2]

Sanpete County has a rich history, not with outlaws and highwaymen, but with a striking comprehensiveness that can't be found anywhere else in Utah. Mixed with mystery, intrigue, and suspense, Sanpete Valley has it all. It was a favorite camping site for the quick-tempered Ute War Chief Walkara, or Walker as he was known by the first white settlers. These first settlers were led by Isaac Morley, who was sent by LDS President Brigham Young to settle Sanpete Valley.[3]

Isaac Morley

The Ute Chief and Morley became good friends[4] after a few scrapes and bruises, but nonetheless they got along very well. There are many stories that surround the pair; some true, some false, but they all seem to fulfill the needs of those who read these tales.

Not to get too deep into this narrative, I will only highlight a few events that I deem relative. First of all, it is generally known throughout the LDS Church membership that Isaac Morley was the "Father" of Sanpete Valley and the Mormons there.[5] His concern was with their welfare and nothing else. He also was instrumental in bringing peace to the Valley with the Utes and became a great friend to all he met.

Unfortunately, there has been much written that cast a strange light upon Morley that has no basis whatsoever. It has been said that he was involved with extracting gold ore from the mountains of eastern Utah with the aid of Ute Indians, such as the brothers of Walkara and, in some cases, with Walkara himself.[6] Those stories, like so many others, have little or no basis of truth. Morley's mission was to establish settlements in Sanpete Valley and to make peace with the Indians and nothing more. He did not have the time or the inclination to go off gold hunting.

ANCIENT ARTIFACTS AND WRITING IN MANTI

Sanpete Valley has seen its share of the strange and unbelievable. Morley was shown an old stone altar very near where the LDS temple now stands in Manti. He said it was the remains of a Nephite altar and the very site where an ancient Nephite temple once stood. He was shown underground tunnels and rooms that once housed ancient Nephite treasure and records.[7] It wasn't by chance that Brigham Young sent Parley P. Pratt to the area first, then Morley to finally settle pioneers there.

When word came that a new LDS temple would be built in Manti, and when its location was revealed, many people set out to see what they could find in and around the site, since it was noted that the ancient Nephites once inhabited the region. Behind the present temple, and where the rock quarry is, some very impressive rock writings were discovered that even today have many scholars and researchers scratching their heads.

Unfortunately, would-be treasure hunters have almost obliterated the left-hand panel of rock writings and have at some extent vandalized the right-hand panel. These are modern-day outlaws who have no regard for ancient sites or private property. Whether these panels were seen by the settlers who came to Sanpete County is unknown since there is nothing written about them. There are some who testified verbally that they were seen by early settlers, but there is no evidence to confirm their claims.

Modern-day outlaws and unscrupulous people have destroyed many ancient sites around Sanpete Valley in an attempt to locate hidden treasures purported to have been secreted away by ancient people. Outlaws don't necessarily have to pack side arms and rob banks and trains; they can use a pick and a shovel or a backhoe to rob and plunder as well.

Manti, Utah, hieroglyphs

NOTES

1. Rufus Wood Leigh, *Five Hundred Utah Place Names: Their Origin and Significance* (Deseret News Press: Salt Lake City, 1961), 86–87.
2. Ibid., 87; italics in original.
3. *Wikipedia*, s.v. "Sanpete County, Utah," last modified March 14, 2017, en.wikipedia.org/wiki/Sanpete_County,_Utah.
4. Kerry Ross Boren and Lisa Lee Boren, *The Utah God Rush: The Lost Rhoades Mine and the Hathenbruck Legacy* (Springville, Utah: Council Press, 2002), 16.
5. *Wikipedia,* s.v. "Isaac Morley," last modified February 16, 2016, en.wikipedia.org/wiki/Isaac_Morley.
6. See Kerry Ross Boren and Lisa Lee Boren, "Chief Walker and the Money-Rock," in *The Utah God Rush: The Lost Rhoades Mine and the Hathenbruck Legacy* (Springville, Utah: Council Press, 2002), 13–18.
7. Kerry Ross Boren and Lisa Lee Boren, *Following the Ark of the Covenant: The Treasure of God* (Springville, Utah: Cedar Fort, Inc., 2000), 116–17.

Sevier County

"Sevier County is one of seven through which the Sevier River flows; by extension, the county was named from that of the river. Sevier County extends from [Sanpete] on the north to Piute and Wayne on the south; and from the crest of Pah Vant Plateau on the west to a south-north line on the east, which is roughly along the east front of Fish Lake Plateau and the southern end of the Wasatch Plateau. . . . The Sevier River courses through the county from the hamlet of Sevier on the south, northeasterly to Redmond; its valley is rich in agriculture and well populated."[1]

"Sevier River has its source in the extreme southeastern part of the Great Basin; its headwaters share the drainage of the High Plateaus of southern Utah with branches of Pahreah River, Kanab Creek, and Río Virgin, all of which . . . flow in the opposite direction into the Colorado River."[2]

Like the Green and Virgin Rivers, the Sevier River's name had its genesis in the original Spanish name *Rio Severo.*[3]

When people mention Sevier County, Fish Lake immediately enters the minds of those who love to camp and fish at this famous lake. Fish Lake has been associated with lost mines, buried treasures, and ancient dolmen and monuments for many years. One famous story that has been told and retold dozens of times and still brings a rush of adrenaline and a quickened heartbeat is the story of Carl Blomquist's lost treasure.

THE LOST TREASURE OF CARL BLOMQUIST

Carl had befriended some local Indians who tried to persuade him to get whiskey for them from the local saloon at Richfield. Carl's father ran the saloon and wouldn't let the Indians enter or buy whiskey. To persuade young Carl further, the Indians appeared one day with some small gold ingots of curious design. The ingots were crudely made and appeared to be very old. They appeared to be a mixture of copper and gold because of the slight green oxidation on them.

After a time, Carl was able to extract a few bottles of whiskey in exchange for the ingots, but had to lie about where he had obtained these small ingots. He didn't want his Indian friends bothered, or even killed, by unscrupulous characters wanting to know where the ingots came from.

After much prodding by Carl, his Indian friends decided to let him in on their secret and show him where they had obtained the little gold ingots. They took him high into the Fish Lake Mountains where they set up camp. A little after sunset, they took Carl to a secreted cave where they showed him a pile of rotten animal hides and pigskin bags. The skins were torn open, allowing some of their contents to spill out onto the floor of the cave. There were many ingots of all sizes, as well as gold nuggets mixed in. Rusted Spanish armor was seen, as well as swords and other items. Carl was not allowed to touch anything but only to look. His Indian friends then told him that he was never to come back to this place or he would be killed. Their ancestors had worked in the cave and many of them died as slaves for the Spaniards. Carl promised he would never seek the cave again, but he didn't keep his word.

Carl went back a number of times, seeking the limestone ledge where the cave was hidden but he never found it. His last trip was by airplane in 1972.[4]

Some say this story is only legend or a tall tale. But is it?

LOST SPANISH MINES

Herbert S. Auerbach, a local historian who was known for his collection of books, manuscripts, documents, and photos of the Old West, mentioned several other tales of treasure in his article, "Old Trails, Old Forts, Old Trappers and Traders: History and Romance of the Old Spanish Trail."[5] One of those tales references a pair of well-known mines, at that time termed "Lost Spanish Mines," namely the "El Dorado" and the "Devil's Palace."

He adds, "Also richly romantic and colorful are the many stories of 'Lost Mines' scattered along the Old Spanish Trail and throughout adjacent territory, many of the legends and traditions dating back to early Spanish explorations and even before that, to mines worked by the Indians. Among these reputedly rich mine along the course of the Old Spanish Trail were . . . the 'Lost Las Vegas Mine' . . . [and] the 'Lost Fish Lake Mine.'"[6]

Other than old Ute Indian legends, there is very little known about the lost mines that the Spaniards worked near Fish Lake. The rock monuments at Fish Lake and those near Aurora, Utah, are copies, or perhaps originals, denoting a method of communication much like the Inuit used in Canada and Alaska. Do they have anything to do with lost mines on Fish Lake Mountain? One thing is certain, during the days when ore was being hauled by teamster Manuel Mestes from Hoyt Peak Summit County to the Gulf of Mexico, a stop was often made somewhere near Fish Lake Mountain.

Pioneer and historian Joseph Fish recorded details of his explorations near Fish Lake Mountain during the summer of 1857:

> We traveled through the upper Bear Valley where I killed the largest rattlesnake I had ever seen! We were following an old trail and probably the one Fremont came in on it 1854. We reached the East Fork of the Sevier River and followed it to Otter creek. There was a great number of antelope but we avoided discharging our guns as much as possible, and we never made a fire for fear of attracting some prowling savage. At about this point we lost the old trail, which we had followed all the way. I have been told that there is an Indian tradition that there was an old gold mine near this place, which was worked by the Spaniards at an early date.[7]

Not only did the Indians and the Spaniards travel the old trails over Fish Lake Mountain, so did outlaws and renegades from all walks of life. From Fish Lake, there are a number of major routes taken by the Spanish to access different parts of the country surrounding Fish Lake Mountain. One trail was down UM Creek where many Spanish symbols have been discovered over the years. Another trail went north toward Sheep Valley on the west side of Mount Hilgard, and then into Red Creek toward Ivie Creek and on into the Cedar Mountain area south of Price, Utah.

The trail was actually many combined into one. The Spanish drove hundreds, if not thousands of mules, cattle, and donkeys over these trails to the north, west, and east for trade and to supply Spanish occupiers with fresh animals and meat. Red Creek hosts several Spanish sites with legible writings on rock and aged old ponderosa trees. Tommy Hollow, just west of Red Creek, was a hideout for outlaws as well. It is assumed that there are outlaws, as well as Spanish caches, buried in and around both Tommy Hollow and Red Creek.

Forest service archaeologist Robert Leonard at the site of a "St. Anthony's Cross" in Red Creek

NOTES

1. Rufus Wood Leigh, *Five Hundred Utah Place Names: Their Origin and Significance* (Deseret News Press: Salt Lake City, 1961), 88–89.
2. Ibid., 89.
3. Ibid.
4. See "Back Reaches of Fish Lake Mtn.," Floyd Mann, *The Treasures of Utah*, forum, accessed April 11, 2017, thetreasuresofutah .yuku.com/topic/998/BACK-REACHES-OF-FISH-LAKE-MTN# .WO06T1MrKit. See also "Spanish Gold Cache of Fish Lake," Igor, *Ancient Lost Treasures*, forum, accessed April 11, 2017, www.ancientlosttreasures .com/forum/viewtopic.php?f=187&t=14217.
5. Herbert S. Auerbach, "Old Trails, Old Forts, Old Trappers and Traders: History and Romance of the Old Spanish Trail," *Utah Historical Quarterly* 9, no. 1 (1941): 13–63.
6. Ibid., 17.
7. Genealogical record, in Lyman D. Platt's possession.

Summit County

"Summit County was created in 1854. Summit was chosen as the name because the county embraces high mountain areas which form the divides between the Weber, Bear, and Green River drainage areas. It includes the crest of the Wasatch, east of Salt Lake and Morgan counties, the Rhodes Valley of the upper Weber River, and from the crest of the High Uintas northward to the Wyoming line and eastward to Daggett County."[1]

MORMON BLOCKADE AT ECHO CANYON

Summit County has a colorful history that is unsurpassed by other counties in Utah. The long list of historical events could fill volumes. We can't talk about Summit County without telling about the historic Mormon blockade of Echo Canyon during those tremulous days of the 1850s.

Associate Justice W. W. Drummond had charged The Church of Jesus Christ of Latter-day Saints with an assortment of crimes ranging from simple treason to mass murder. In Drummond's resignation letter, read aloud in Salt Lake City, the first news of an impending attack came to the Saints. They read in disbelief what Drummond had so callously written:

> In the first place, Brigham Young, the Governor of Utah Territory, is the acknowledged head of the "Church of Jesus Christ of Latter-day Saints," commonly called "Mormons," and, as such head, the Mormons look to him and to *him alone* for the *law* by which they are to be governed:

> therefore no law of Congress is by them considered binding in any manner.
>
> Secondly. I know that there is a secret oath-bound organization among all the male members of the Church to resist the laws of the country, and to acknowledge no law save the law of the "Holy Priesthood," which comes to the people through Brigham Young direct from God; he, Young being the vicegerent of God and prophet, viz: successor of Joseph Smith, who was the founder of this blind and treasonable organization.
>
> Thirdly. I am fully aware that there is a set of men set apart by special order of the Church, to take both the lives and property of persons who may question the authority of the Church; the names of whom I will promptly make known at a future time.
>
> Fourthly. That the records, papers, etc., of the Supreme Court have been destroyed by order of the Church, with the direct knowledge and approbation of Governor B. Young. . . .
>
> Sixthly. That the Federal officers are daily compelled to hear the form of the American government traduced, the chief executives of the Nation, both living and dead, slandered and abused from the masses, as well as from all the leading members of the Church in the most vulgar, loathsome and wicked manner that the evil passions of men can conceive. . . .
>
> That Captain John W. Gunnison, and his party of eight others, were murdered by the Indians in 1853, under the orders, advice and direction of the Mormons; that my illustrious and distinguished predecessor, Hon. Leonidas Shaver, came to his death by drinking poisoned liquors given to him under the order of the leading men of the Mormon Church in Great Salt Lake City; that the late Secretary of the Territory, A. W. Babbitt, was murdered on the plains by a band of Mormon marauders, under the particular and special order of Brigham Young, Heber C. Kimball, and J. M. Grant, and not by the Indians, as reported by the Mormons themselves, and that they were sent from Great Salt Lake City for that purpose and for that only; and as members of the Danite band they were bound to do the will of Brigham Young as the head of the Church, or forfeit their own lives.[2]

Shortly after, President James Buchanan ordered an army to Utah, known as the Utah Expedition. Echo Canyon became the first obstacle the army had to cross before reaching the Salt Lake Valley. History tells us that Brigham Young moved his people out of the city in preparation for the invading army.[3]

Even today, stone fortresses can be seen on the ledges of the red rock high above the railroad tracks. People still investigate these old ruins and sometimes make little discoveries, such as bullets, buckles, or even a pocket watch.

Echo Canyon was home to the Needle Rock Station on the old Pony Express and Overland Stage Trail.[4] It was built near a prominent landmark of rocks, which were used as a guide by the early pioneers. This is a favorite among metal detector enthusiasts.

CACHE CAVE

Cache Cave was used by Indians, explorers, and fur traders, and later became a stage station on the Old Overland Stage Route. During the Utah War, Mormon leaders used it as a headquarters.

The site was also used by the notorious Potter outlaw gang. It was rumored that Ike Potter and Charley Wilson often buried stolen loot and treasures at or near Cache Cave.[5] In 1867, the gang was killed in a daring jailbreak at Coalville. Their plunder was never recovered.

Cache Cave

WEBER STATION

Weber Station at the mouth of Echo Canyon stood for many years. When it was abandoned, it sat in a dilapidated state until 1931 when it was torn down. Numerous relics from the Pony Express era including several $5 gold pieces were discovered. During the construction of the railroad down Echo Canyon in 1868, tent saloons, gambling tents, brothels, and more sprang up to rob the hard-working railroaders from their pay. It was common to see a man one day, then never see him again. Many mysterious disappearances seemed commonplace. Seven bodies were found under one saloon alone, lending credence to the violence. The site was used as a gas station until the early 1930s when it was torn down to reveal their gruesome deaths.[6]

The notorious Rachet outlaw gang used Weber Station as a headquarters during the 1800s. Some say they buried some of their booty either at the station or nearby.[7]

TREASURE IN ECHO CANYON

Not only did the outlaws rob the folks in and around the canyon, but so did nature herself. Just west of what was called in those days Cobblestone Hill in Echo Canyon, right off the old stagecoach route, was a very nasty curve in the road. If a teamster wasn't on his toes and had a strong command of his horses, disaster was sure to follow. Unfortunately, a few accidents took place, mostly in the winter, which found coaches, horses, passengers, and their belongings down over the edge, if not clear to the bottom of the canyon. If one could access that site with a metal detector, more than likely artifacts and relics could be recovered.

Echo Canyon and its creepy little hamlets were boom places for the worldly and profane. No store ever locked its doors. Every saloon had people drinking in them at any hour of any day. Robberies, fights, and shootings were not uncommon. Of interest to treasure hunters is the fact that these little hamlets had more than their share of tough guys, prostitutes, and shysters who spent, robbed, and no doubt lost many a gold coin or bit of gold dust through the cracks of the old wooden floors. A well-known story tells us that years after Echo City was abandoned, a young farmer, wanting some lumber, tore down a room in the old rock jail. Behind some loose rock, he discovered a gold watch, a pair of reading glasses, and a bag of gold coins.[8]

BURIED LOOT AT ECHO CANYON

Stories abound in and around the area such as the one of two outlaws who "robbed a stagecoach in the 1860s at Echo Canyon. The outlaws were captured at their hideout up in the mountains, one or two hours hike from an old ranch that was located between six to nine miles east of Weber Station. The loot was buried in the area of their camp and never recovered. One of the bandits was killed in an escape attempt and the other was sentenced to prison where he died."[9] Most of the land is now private, so if you want to search for treasure, please contact the landowners.

COALVILLE'S MOST FEARED OUTLAW

One man who terrorized the people of northern Utah was from their rank and file membership was Isaac Smith Potter.

Isaac Smith Potter's life began at Loraine County, Illinois, in 1833. His parents converted to the LDS faith and moved west to Springville, Utah, in 1852. He and his father ran a farm, raising cows and corn. There was a band of Indians who lived near the Potter farm, and their young men became good friends with "Ike" Potter. Potter didn't stay to the true blue life of a Latter-day Saint; instead, he took up an opposite way of life, not in harmony with the Church's teachings.

Court records found at Summit County tell a different story about Isaac "Ike" Potter. This Potter was not the sweet-loving boy who worked so diligently on his father's farm, but was a man who "struck terror in the hearts of hardened pioneers in Northern Utah in the early [1860s], far more than a gun-toting outlaw could have done."[10] The threat of Potter's raid was on the families of Summit County. Families were not safe; heavy armed guards patrolled the town of Coalville, and these patrols continued until "Isaac Potter and Charley Wilson were shot to death, and John Walker was wounded in their attempt to escape guards the night of August 1, 1867."[11]

> Several events led to the shooting of Potter and Wilson. On June 10, 1867, Potter was given 4500 pounds of grain to deliver to Fort Bridger, Wyoming. Instead, he hid the grain in Cache Cave. Action was brought and he was found guilty.
>
> In July 1867, a group of men, working for the old Springville Mining Company, reported being stopped on the way home by Potter, Wilson, and Walker. Potter said they were going to Fort Bridger for help, as they were being persecuted by the people of Summit County. If help could not be obtained from the Fort, they would call on Black Hawk (the famous [Ute] war chief) and attack Coalville.
>
> William Cluff, mayor of Coalville, immediately placed a heavy guard around Coalville. Afterwards, he testified in court he had counted ten white men besides a few Indians in Potter's band.
>
> Unable to influence the soldiers at Fort Bridger, Potter and his gang came back, encountering a band of Indians near Bear River. The Indians, however, refused to follow Potter on a raid. Potter then returned to Summit County with his two men.

> Mr. S. James Bromley, owner of the famous old Weber Stage and Pony Station [near Echo, Utah] and a great friend of the Indians, reported general unrest among them.
>
> On August 1, 1867, [George] C. Roundy, sheriff of Summit County, was ordered to arrest Potter and his men, on a charge of stealing cattle. In the round-up, Isaac Potter, Charley Wilson and John Walker were captured and placed in the custody of Alma Eldredge, Deputy Sheriff. Eldredge placed the three men under guard of two special deputies, Joshua Wiseman and James Mahoney.
>
> That night the three prisoners made an attempt to escape, thinking Mahoney had gone home and Wiseman was the only guard on duty. Wiseman was sitting near the door reading, and was knocked to the floor. Mahoney, standing in the yard, saw the men as they ran from the building and ordered them to stop. Seeing they did not intend to stop he opened fire. At the same time Wiseman fired from inside the jail. Potter was killed a short distance from the jail, Wilson was not found until morning. Laying with his hand and part of his body underwater, in Chalk Creek, three hundred yards from the jail. Walker when located, was so seriously wounded, he could not be brought to testify in the investigation against the sheriff's force, on charges of murdering Potter and Wilson. It is not known whether Walker died of his wounds or recovered.[12]

One thing that has stumped researchers, as well as the people of those days, is what was told to them by the two guards: Wiseman and Mahoney. They claimed they were so excited they didn't know if their shots took effect. Wiseman stated that while in pursuit of Wilson, he was nearly killed by bullets flying around him.

It's possible that, based on their testimonies, neither Wiseman nor Mahoney shot the prisoners, but were aided in the shooting by parties unknown to them.

METAL DETECTING IN SUMMIT COUNTY

Summit County is also known for its rich mining history and tales of lost Spanish gold. Stories of lost mines and mule trains of Spanish gold coming out of the western Uinta Mountains would fill volumes. It's not hard to find information on just about every aspect of these tales, but finding the correct and fact-filled info is. Most of the written stories, save a few, have been embellished and have taken on the air of a Hollywood movie. Nevertheless, the stories bear some truth and are loved by many who venture out into those wild mountains, searching for the mines and treasures they believe were left behind by Spanish miners.

To keep those stories alive, many artifacts have been discovered from the time of the first Mormon settlers to modern-day prospectors. Stories of "I saw it but lost it again" are normal talk among treasure hunters. Before the advent of the GPS, only maps and compasses were used. Now, finding something in the wilderness and keeping track of it shouldn't be a problem, but believe it or not, "I marked it but couldn't find it again" is still heard.

Having a good metal detector is a must if you are going to hunt for treasure, not to mention you'll need some reliable information you can count on. Family outings are great when hunting for old mines and treasure caches because it's something the whole family can enjoy. Just getting into the great outdoors is exciting enough for many folks.

THE RHOADES MINES

The story that seems to entice the prospecting blood in all of us comes from the history written by Gale R. Rhoades concerning Thomas and Caleb Rhoades and the Spanish mines they removed gold from. Gold was discovered east of Rhoades Valley early on and was used as trade by the local Indians for goods and foodstuffs. Spanish mule trains were seen coming out of Beaver Creek, Weber River, and the South Fork of the Provo.

Brigham Young noticed the gold trinkets and ornaments worn or traded by the Ute people and made an inquiry as to where the "money rock" came from. He became friends with Chief Walkara and, in time, the chief told Brigham about the mines east of Salt Lake City. Despite the fact that it was equally as irreligious for an Indian to show a white man the source of their sacred gold as it was for the Mormons to mine it, both the Utes and the Mormons found an escape in their beliefs. They entered into an agreement that was unique throughout the West because it was the only time in recorded history that the Indians willingly revealed to a white man the location of their sacred and fantastically rich gold mine, and that the white man did not reveal its location to any man during his entire lifetime.[13]

And so it was that in July of 1852, an agreement was made between Chiefs Walkara and Soweet of the Ute Tribe and Brigham Young and Thomas Rhoades, to the effect that Walkara would give to the Church all the gold that it needed if the origin of the gold deposit was kept secret. Walkara "set down three specific conditions that were to be strictly adhered to before Thomas Rhoades could enter the mountains for the purpose of obtaining the precious gold."[14]

Condition (1) "The location of the gold would be revealed only to one man, mutually trusted by Brigham Young and the Utes; this man would be kept under constant surveillance at all times while in the area and Brigham Young himself, for purposes of assurance, would not know the location of the gold."[15]

Condition (2) "The penalty of death would automatically be incurred upon the man chosen to retrieve the gold if he gave the secret of its location to any man or attempted to bring out gold without the permission of the Utes, expressly Walker or the head chiefs, though he could bring out as much of the precious metal as he could pack for the use of the Church. Any white man who attempted to follow the chosen man to the gold's location would also suffer the death penalty."[16]

Condition (3) "Walker expressed the fact that the Indians would not, at any time, assist in the mining of the gold, since no large scale operation was necessary anyway and they needed only little for their own use. Besides, the Utes were hunters, not miners,and they were still seething from the forced labor pushed upon them only years earlier by the Spaniards."[17]

Hunting for the lost mines east of Kamas has been an ongoing event for some time. Many think they have found those lost mines, but miners from the days of the marauding Spaniards dug holes all over the western Uinta Mountains. Some mines turned out fairly well while most were nothing short of a waste of manpower, abandoned for a richer-looking area.

THE LOST JOSEPHINE MINE

Other mines and lost treasures include the Black Bull Cache, Big Elk Lake Mine, Gardner Fork Mine, Mine of the Pass, Norway Flat Mine, Iron Mine Mountain Cache and Mine, Notch Mountain Mine, Mount Watson Mine, Hayden Peak Mine, Dead Man Mountain Mine, and the famous Josephine Mine.

There are many mines named Josephine, but the *real* Josephine Mine is in Summit County. As stated before, in the chapter about Garfield County, although there are three sites in Utah that claim to be the location of this famous mine, only Hoyt Peak in Summit County is actually documented as such.

FINDS IN KAMAS

The area north of the Provo River, south of the Weber River, west from Mirror Lake, and east from Kamas Valley has produced some interesting artifacts dating as far back as AD 253.

One artifact is in the form of a Roman coin, discovered by a man named Lannie Seely of Salt Lake City while prospecting in a mine on Hoyt Peak just east of Kamas. How the coin got to that place is still a mystery. These few photos below give a peek at what has been discovered.

There are numerous artifacts that have been discovered in Kamas that have been written about in other publications, and a few that have been kept secret for obvious reasons. The area mentioned above is a great place to metal detect and to explore. It never gets boring or mundane and can be therapeutic to the soul.

"THE GREATEST SILVER MINE IN THE WORLD"

Park City is undoubtedly known best for the Sundance Film Festival held every year during the month of January. Skiing is another great enjoyment had by the people that visit the town every year.

Founded in 1870, the city started as a mining boomtown. The old abandoned Ontario Mine, located in the hills west of the city, produced around $14 million and was dubbed the "greatest silver mine in the world."[18] When the mine began playing out in the late 1880s, the Chinese population soared to work it.

Roman Antoninianus coins

Back piece. At the bottom of the back piece are engravings that suggest it was most likely German in origin.

Sword found in Kamas

NOTES

1. Rufus Wood Leigh, *Five Hundred Utah Place Names: Their Origin and Significance* (Deseret News Press: Salt Lake City, 1961), 94.
2. Junius F. Wells, ed., *The Contributor: Representing the Young Men's and Young Ladies' Mutual Improvement Associations of the Latter-day Saints* (Salt Lake City: The Contributor Company, 1888), 9:245–46; italics in original.
3. *Wikipedia*, s.v. "Utah War," last modified April 6, 2017, en.wikipedia .org/wiki/Utah_War.
4. "Utah Pony Express Stations," *Expedition Utah*, accessed April 11, 2017, www.expeditionutah.com/featured-trails/pony-express-trail/utah-pony -express-stations/.
5. "Needle Rock Station and the Cache Cave Treasure," *Utah Education Network*, accessed April 11, 2017, www.uen.org/Centennial/19GhostsB02 .shtml.
6. Ron, "Utah Ghost towns, relics and gold," *Treasure Hunting*, okietreasurehunter.blogspot.com/2009/10/utah-ghost-towns-relics-and -gold.html.
7. Ibid.

8. "Treasures from the Pony Express Trail," *Treasure Hunter Depot*, accessed April 11, 2017, www.hunterdepot.com/treasure-stories-legends/pony-express-trail/.
9. "Rich County," Floyd Mann, *The Treasures of Utah,* forum, accessed April 10, 2017, thetreasuresofutah.yuku.com/topic/974/RICH-COUNTY#.WOwIm1MrKis.
10. "Utah's Terror," *Family Search*, accessed April 11, 2017, familysearch.org/photos/artifacts/6810403.
11. Ibid.
12. Ibid.
13. See Twila Van Leer, "Rhoads, Gold Seemed to God Hand in Hand," *Deseret News*, July 2, 1996, www.deseretnews.com/article/499383/.
14. Gale R. Rhoades, *Footprints in the Wilderness: A History of the Lost Rhoades Mines* (Dream Garden Press, 1980), 113.
15. Ibid.
16. Ibid.
17. Ibid.
18. "Fun Facts about Park City," Patrick Howell, *Mountain Home Team—Park City Utah*, accessed April 11, 2017, mtnhomeparkcity.com/fun-facts-about-park-city/.

Tooele County

"Tooele County was one of the original eight created in 1850. Tooele is one of Utah's largest counties: it embraces the major part of Great Salt Lake Desert, southwest part of Great Salt Lake, Skull and Tooele valleys; it extends from Nevada eastward to the crest of the Oquirrh Range on the west; and from Box Elder on the north to Juab County on the south. Tooele County was named from Tooele Valley."[1]

> In Gunnison's *History of the Mormons* is the following: "In Tuilla valley, thirty miles west of the temple, is a settlement." *Tule*, a word of Aztec origin, is a large variety of the common great marshes of the valley today. Gudde writes that "*tule* derives from the Aztec *tullin* or *tollin*; the word designates the bulrush or similar plants with sword-like leaves, as shown in the Aztec symbol. In Spanish exploratory times the word *tule* was used descriptively." . . . The Spanish pronunciation of *tule* (too'-le) is suggestive of the spelling as given by Gunnison and others then current. *Tuilla* could not be derived from the Gosiute (Shoshoni) word for tule, *saip*. It is related that Brigham Young's scribe, not knowing the current orthography of the valley's name, *Tuilla*, spelled it as it sounded [phonetically]: *Tooele*, which corruption has since been perpetuated. Thus, the name is traceable back through Mormon and Mexican corruptions to its ultimate Aztec origin.[2]

When members of the army camped at Camp Floyd and Camp Douglas, they discovered gold and silver in the Oquirrh Mountains, and a new beginning took place. It forever changed the face of Tooele County and launched the county into a high fever of expectations by speculators and outlaws.

GOLD CANNONBALLS AT GOLD HILL

Wally Buhechcer lived in Salt Lake City, but loved to venture into the Deep Creek Mountains in western Tooele County, looking for what he said was a lost treasure trove hidden by army deserters from Camp Floyd. He and his daughter, Leslie, would travel to the mountains and camp for weeks on end, hunting for this elusive cache. Soon, however, his daughter married and that ended his prospecting and treasure-hunting partnership with her.

One day, while out wandering through the tall sagebrush, not far from the town of Gold Hill, Wally stumbled upon a neatly stacked pile of cannon balls. He knew right then he had found the lost loot. Tossing the cannon balls into the surrounding sage, he cleared the spot beneath and began digging. However, it was not to be, for there was nothing beneath but dirt.

Wally decided to take a few of the cannonballs to Salt Lake City for testing and to find out if in fact they were army-style balls. In time, he found out that they were not army issue but more than likely were Spanish. This shed new light on things as far as Wally was concerned. Had his resources and connections been flawed? Was it really a Spanish cache he was looking for and not an army stash?

He began to research his information again and soon learned that, in fact, it was a Spanish cache that the deserters had discovered and reburied. He found a map and a short note that described the discovery. His source was an elderly woman who was a granddaughter of one of the deserters. How he came to know this woman he never would say, but somehow he gained her confidence and ended up with the note and the map.

Wally spent most of his retired years in Gold Hill, searching and hunting. His map took him to what he called "Blood Canyon," where he spent a lot of time digging and hunting for a clue that would give him the cache he coveted.

One day, near an old spring in Blood Canyon, he discovered a few very heavy, odd-shaped iron balls. Wally took the balls to a friend of his in Salt Lake City and had him cut one of them in half using a diamond rotary saw. To their amazement, they saw the center of the iron ball was filled with gold. Someone had found a way to cast iron with a hollow core, then pour molten gold into its center. How clever!

I met Wally in his later years and spent time interviewing him about his discovery. Wally said that the iron and gold balls were Spanish, and that by filling the iron balls with gold, the Spaniards could transport them without the Indians knowing what they were up to.

Wally claimed that he had a Goshute friend who gave him information on the Spanish and their mining activities in the Deep Creek Mountains. He said that more of those iron and gold balls were still hidden in spots within sight of Blood Mountain. Wally had done well with the gold he'd already discovered, but was very tight-lipped as to giving out too much information about the gold still hidden.

Not long after the interview, Wally contacted me. He gave up his map and notes to me, on the condition that if anything was ever found, his posterity would benefit. It is my belief that there are two caches: the army cache and the Spanish iron and gold balls cache.[3]

THE MORMON MINT

In 1848, the newly established settlement of Salt Lake City was in dire need of many things, including money to assure the health of communities. The people developed a Mormon script—a form of money that could only be redeemed at the Church's own businesses, but coinage seemed more practical and much more desirable. President Brigham Young desired that a mint should be established and gold coins should be made available. Gold brought in from California and from sources like the Rhoades's Mine would be used for this coinage undertaking.

A traveler who was passing through Salt Lake City heard of the Mormon mint and that the mint itself was poorly protected. This traveler apparently waited for the cover of darkness and then snuck up to the mint, where he discovered a pile of freshly minted coins. He gathered up what he could and got away as fast as he could.

This man was not alone in the heist but had someone holding the horses nearby. It was discovered that the men were two brothers who had been attached to a wagon train traveling to California. They were identified as the Baldwin brothers, and that gave law enforcement an edge on whom to look for.

Unfortunately, by the time Porter Rockwell, the famed "Mormon marshal," caught up with the wagon train, he discovered that the brothers, upon hearing of a posse headed their way, had gone into the vast western desert. It was figured the two got away with at least two hundred coins, all of which were $10 gold pieces. Today's value would be astounding.

It is not known if the two men made their way to California or died in the desert.[4] One story has it that, as late as 1970, the town marshal of Ophir made a unique discovery while riding his horse in the west desert region of Utah. His discovery included the bleached bones of a human with saddlebags, a rifle, and a saddle.[5] The bags were full of letters and documents.

It was decided the remains were those of a Pony Express rider who had disappeared in 1861. However, some folks believe that Porter Rockwell caught up with the Baldwin brothers, killed them, and then buried the gold coins somewhere in Government Creek near his ranch. Of course this is all hearsay.

CACHE AT DUGWAY MOUNTAIN RANGE

In 1854, a group of California miners ran out of provisions while out in the west desert. They needed to acquire food soon or perish. It was decided by the group to take as little as possible and go in search for food. They buried what they didn't need, as to not weigh down their animals with unnecessary items, then headed toward Salt Lake City. The items buried added up to about $70,000 in gold coin and ingots.

They were lucky to come upon a western-bound group of travelers that sold them enough provisions to last them until they could get to a permanent settlement. Upon their return to what is now known as the Dugway Mountain Range, they tried in vain to locate the site where they had hid their cache of gold and other possessions. To their utter dismay, they never relocated it.

The only clue they left behind was that the cache site was close by a well or spring of water.[6]

"DUTCH" SLOSINGER'S MINE

There are many stories of gold, silver, and hidden and lost mines in Tooele County. For instance, in 1880, an old German prospector known as "Dutch" Slosinger discovered a very rich gold mine near Dutch Mountain, just north of Gold Hill. He mined enough gold to get him back to Germany and to live comfortably once there. He left a map and directions with a friend who lived in Salt Lake City, but his friend was never able to locate the mine.

OLD JOHN'S CACHE

"A flask containing one hundred pounds of gold was buried somewhere around the cabin owned by a miner known only as 'Old John' in the 1940s at a ghost town Gold Hill."[7] His cache was never recovered.

RUMORS OF HIDDEN LOOT

Another cache of nearly half a million dollars is rumored to be hidden somewhere on the edge of the Great Salt Lake Desert. It is believed that it was a favorite hiding spot for loot robbed off of stages and overland immigration trains.[8] Of course there is no way of really knowing whether this is true or not.

PORTER ROCKWELL'S ORE CLAIMS

"An outlaw robbed the Overland Stage of $40,000 in gold coins between Dugway and Riverbed Station. Porter Rockwell, famed Mormon marshal, followed the tracks to his camp near the south end of the East Tintic Mountains, arrested the man and recovered $30,000 from several shallow holes where the outlaw had buried it. He unknowingly missed $10,000 of the treasure and it has never been recovered. The outlaw was killed while trying to escape shortly afterwards."[9]

Porter Rockwell

Some say Porter killed him because he wanted to keep the $10,000 for himself and didn't want the outlaw to make waves. Of course, this theory doesn't make sense if you take Porter's word that he unknowingly missed that much of the treasure.

Others say the cache is still out there because Porter would never stoop so low as to steal from his Church or from his peers.

Porter Rockwell was fond of gold and silver and had several claims throughout the region. Rockwell hated the fact that Colonel Patrick Connor's men had permission to exploit Utah's hidden mineral wealth.

The officer's efforts through letters and newspaper articles in the East eventually influenced hordes of prospectors to try their luck in the territory and gained for Connor the reputation of being the "Father of Utah Mining."[10]

The last thing Porter wanted was some stranger snooping around the mountains in search of gold and silver. In a sermon delivered June 17, 1877, in Farmington, Brigham Young recounted Porter's apprehensions that his mine would be found. (According to Porter's descendants, he had a large cache hidden there.)

Young recorded:

> Porter, as we generally call him, came to me one day, saying, "They have struck within four inches of my lode, what shall I do?" He was carried away with the idea that he must do something. I therefore told him to go with the other brethren interested, and make his claim. When he got through talking, I said to him, "Porter, you ought to know better; you have seen and heard things which I have not, and are a man of long experience in this Church. I want to tell you one thing; they may strike within four inches of that lode as many times as they have a mind to, and they will not find it." They hunted and hunted, hundreds of them did; and I had the pleasure of laughing at him a little, for when he went there again, he could not find it himself.[11]

NOTES

1. Rufus Wood Leigh, *Five Hundred Utah Place Names: Their Origin and Significance* (Deseret News Press: Salt Lake City, 1961), 96.
2. Ibid., 96–97; italics in original.
3. Wally Buhechcer, interview by Stephan Shaffer.
4. "Lost Mormon $10 Gold Coins," *Treasure House Relics Project,* accessed April 12, 2017, www.miningutah.com/id387.html. See also "Mormon Mint," *Utah Treasure,* accessed April 12, 2017, utahtreasure.blogspot .com/2010/12/mormon-mint.html; "Baldwin Brothers and the Stolen $10 Mormon Gold Coins," WheatbackDigger, *Treasurenet,* accessed April 12, 2017, www.treasurenet.com/forums/treasure-leads/117676-baldwin -brothers-stolen-10-mormon-gold-coins.html; and "Mormon Gold Mint Robbery," Cleetus, *Treasurenet,* accessed April 12, 2017, www.treasurenet .com/forums/treasure-leads/266595-mormon-gold-mint-robbery.html.
5. "Lost Mormon $10 Gold Coins," *Treasure House Relics Project,* accessed April 12, 2017, www.miningutah.com/id387.html.
6. Robert F. Marx, *Buried Treasure You Can Find* (Dallas Texas: Ram Publishing, 1993), 341.
7. "Gold Hill," Floyd Mann, *The Treasures of Utah,* forum, accessed April 10, 2017, thetreasuresofutah.yuku.com/topic/946/GOLD-HILL# .WOwFPVMrKis.
8. See Robert F. Marx, *Buried Treasure You Can Find* (Dallas Texas: Ram Publishing, 1993), 341.
9. "Riverbed Station," Floyd Mann, *The Treasures of Utah,* forum, accessed April 10, 2017, thetreasuresofutah.yuku.com/topic/939/RIVERBED -STATION#.WOwFQVMrKis.
10. "Colonel Connor Filled a Varied, Dramatic Role in Utah," Jeffrey D. Nichols, *Utah History to Go,* accessed April 12, 2017, historytogo.utah.gov/utah_chapters/mining_and_railroads /colonelconnorfilledavarieddramaticroleinutah.html.
11. Brigham Young, in *Journal of Discourses,* 19:37.

Uintah County

"Uintah County was one of the original eight organized in 1850; it then embraced the areas now included in Duchesne and Daggett counties. It is a long rectangular domain extending southward from the crest of the Uinta Mountains to a common boundary with Grand, in the eastern tier of counties. The county was named for a division of the Ute Indians. *Uintah*, a variant of *Uinta*, is applied to political entities, whereas *Uinta* is applied to natural features and to the *Uinta Utes*. The form *Uinta* is a contraction of *Uintats*, a division of the Ute living in northeastern Utah."[1]

Uintah County has a very colorful history dating back two hundred years before the arrival of the Mormons. The Spanish made an effort to colonize and civilize, but in reality they were there to conquer the Utes.

As early as the sixteenth century, Spanish mining activity took place all around the Vernal and Roosevelt areas. Even in their own records, they mentioned mines they found that predated them by many years, suggesting another group of gold and silver hunters once inhabited the region. Ute historians have said that there was a race of people that came many years ago that had great knowledge of mathematics, farming techniques, and mining. Strange and ancient symbols have been discovered in and around Vernal that lends credence to their claims.

Some of this evidence can be seen northwest of Vernal in Dry Fork Canyon and areas close by. On Sadie McConkie's ranch, some very human forms can be seen that depict a race of people other than Ute or like tribes of the region; however, other human forms do represent tribes such as

Fremont and Ute.[2] Strange symbols such as those shown below can only be found in Dry Fork Canyon.

Many stories abound of Spanish silver mines and the Lost Rhoades Mines being in the area. Many have searched but none have found those elusive mines and treasures.

Infamous outlaws seemed to make Uintah County a place for retreat and hiding out from the law. One such hideout was the little burg of Bullionville, twenty miles north of Vernal.

Founded in 1880 as a mining town, Bullionville was a hangout for such noted outlaws like Matt Warner, Eliza Lay, and Butch Cassidy.[3] There are numerous rumors of outlaw loot buried in the immediate area of Bullionville.

Strange symbols in Dry Fork Canyon

Unfortunately, there is nothing left of the old town. The old board homes and fences have gone the way of the campfire. The boardwalk survived longer than the town itself, but it too became fire fodder.

Hundreds of campers, hunters, and metal-detector enthusiasts have combed the old town from one end to the other. Some have made great discoveries there, but most come away with nothing but a good time.

There is treasure there, but the question is *where*? More than likely, a treasure would be buried somewhere outside of town but still within visual range of the owner of the treasure, or by a landmark of some kind. However, over the years, everything seems to change, so finding such a cache would be difficult, if not impossible.

TABBY WEEP WHITE

Tabby Weep White was a Ute gunfighter with a reputation of being so mean spirited that he would do anything to start a fight. Tabby frequented what was known as the Duchesne Strip, later known as Moffat. He was noted as the "fastest, deadliest man with a pistol in the Territory of Utah," with maybe one exception, that being Harvey Logan of the Wild Bunch. Most could not match Tabby's speed and unerring accuracy with his Colt 44.[4]

"One day Tabby shot it out with three opponents in front of the general store. He was hit twice but walked away, alive, leaving the dead behind."[5]

Another time, while in a saloon, he shot a man simply because he missed the spittoon with his wad of chewing tobacco.

There are many stories and versions of Tabby's shootouts in front of the general store and at the saloon and, as time progressed, the stories became even more embellished.

Truth be told, old Tabby was sent to prison for eight years for the general store killings. After his release, he refused to carry a gun and that would prove to be a critical mistake for Tabby. He was shot in the back of the head while playing poker. His killer was one of the family members of a man he had killed. Tabby had a cache hidden somewhere near the old town of Randlett, Utah. It is said to have gold coins. It may very well be off reservation land. I have the landmark used by him given to me by the late Bishop Arrowchis, Holy Man for the Ute Tribe.

NOTES

1. Rufus Wood Leigh, *Five Hundred Utah Place Names: Their Origin and Significance* (Deseret News Press: Salt Lake City, 1961), 100; italics in original.
2. "Great Escapes: Petroglyphs in Dry Fork Canyon, Utah," Denise Seith, *RV Life Magazine*, accessed April 12, 2017, rvlife.com/330-2/.
3. Peter Massey and Jeanne Wilson, *Backcountry Adventures Utah: The Ultimate Guide to the Utah Backcountry for Anyone with a Sport Utility Vehicle* (Adler Publishing, 2006), 27.
4. "Most Lawless Town in Utah," *Salt Lake Tribune*, September 29, 1968.
5. Ibid.

Utah County

Utah County and Utah Valley are filled with historical events. The early Spanish gold expeditions saw the valley as a temporary home while hauling mineral wealth from the rich Uinta Mountains to Santa Fe, New Mexico. Later, the Dominguez-Escalante expedition found its way to Utah Valley, "attempting to find a way over the *tierra incognita* between Santa Fé . . . and Monterey on the coast of *Alta California*."[1]

The members of the Dominguez-Escalante expedition have historically been given credit for naming the Ute people of Utah Valley *Yutahs* while they were "encamped on the shores of Utah Lake" in 1776.[2] However, history tells us the name *Yutahs* was the name Francisco de Ibarra used when he described his expedition north in 1565.

Many more expeditions found their way north after Ibarra, such as Fray Bartolome Romero and Fray Francisco Muniz in 1604, and Maestre de Campo and Vincent de Saldivar in 1618. Campo and Saldivar were also accompanied by Father Lazarus Ximenez of the Order of St. Francis.

THE SLAVE TRADE IN UTAH

It was well-known among the settlers of Utah Valley that not only did the Utes believe in the slave trade, but so also did the Spanish and Mexican miners that often passed through the country on their way to and from the gold and silver mines east of the valley.

One of the first actions taken by Brigham Young was to prohibit slavery in 1852. Indians could no longer own slaves from other tribes. Later that same year, Mormon V. Selman reported that the gold miners camped

A typical Spanish expedition

on his farm had slaves with them. Marshal William H. Kimball arrested the party. The Mexicans (as he called them) reluctantly surrendered the Indian children and women they had captured, telling the marshal that slaving was more lucrative than mining gold.[3]

Settlers in Utah Valley were warned by Brigham Young not to become infatuated with the idea of getting rich by hunting gold but to stay home and farm. However, a few men hungry for quick riches didn't heed this prophetic warning.[4]

THE MORMON GOLD RUSH

Settlers heard of a lost, centuries-old Ute Indian silver and lead mine on the western slopes of Mount Timpanogos. Disregarding Young's counsel, a few men went hunting for this lost mine. However, greed overtook the men and a gunfight broke out. Nobody was killed, but the event started a feud that was like cancer, and it spread throughout the Mormon communities anywhere gold, silver, or other metals were thought to be present.[5]

LOPEZ'S HOARD

Juan Lopez joined the Ike Potter gang in the 1860s and soon came into possession of a sizeable hoard of Spanish gold coins. The government decided they wanted those coins and so a detachment of soldiers was sent to get Lopez "dead or alive."

Lopez's hideout was in the mountains on the backside of Mount Timpanogos, about where the Sundance ski lodge is today. He was driven from his camp down into American Fork Canyon, where he became trapped and was shot to death by troopers. No gold was found on his body, but when he left his hideout, the troopers said he was packing heavy.

Nobody knows how much gold dust, nuggets, or ingots were lost along the way. Perhaps it's worth the price of a good metal detector to find out.[6]

MILITARY CACHE IN SPRINGVILLE CANYON

> On 9 November 1858 [near the town of Fairfield], amid gunfire and patriotic music, the soldiers of Camp Floyd . . . raised the United States flag above their newly completed garrison. Named for Secretary of War, John B. Floyd, the post housed the largest concentration of U.S. troops to that time, in what immediately became the third largest city in Utah.
>
> Camp Floyd was a product of the so-called "Utah War." Influenced by rumors of rebellion in Utah, President James Buchanan ordered 2,500 soldiers led by General William S. Harney to the territory in May 1857. Colonel Albert Sidney Johnston took over the command of the Utah Expedition as Harney was retained in Kansas to direct troops in the escalating troubles there.[7]

When the Civil War broke out, the army mustered all of its troops at Camp Floyd to head east to engage in the war. Camp Floyd was abandoned, but before the troops left, they buried hoards of military supplies where they figured the Mormon people would never find it. The cache was said to consist of over 2,000 rifles, a large quantity of pistols, swords, and other gear. Some say it was buried within the vicinity of Fairfield and Cedar Fort. It was not. It was taken to Springville Canyon and hid up in a cave.[8]

In 1993, a man by the name of Ray Anderson came forward with an astounding story of a military stash he and a friend found in Springville Canyon. Anderson had previously met a woman in California who for reasons unknown related the following story to him:

> She was the granddaughter of an old rancher who lived in Springville Canyon, on the Left Fork. Just before he died, he related to her that he had been scouting for lost cows in the thick brush and timber on the north side of the road when he came to what appeared to be a well-hidden mine entrance. He would have never seen it if he hadn't stood right in front of it.
>
> Curiosity overtook him, so he planned on returning to see what this mine was all about. Weeks later, he did return with ropes and lights. The entrance was a wooden door that had been concealed with rock and dirt.

> Over time, the wooden door had rotted, so getting past it wasn't that difficult. Once inside, he was amazed at what he had discovered. This large cave was not a mine, but a major cache of military equipment that looked like it was from the Civil War era.
>
> He counted four cannons, several cases of rifles and pistols, and kegs of black power. He thought at that time the army was aware of this cache site so he didn't think it wise to take anything out of the cave. He noticed a shaft coming down from above that looked like it might have been an airshaft. Perhaps it was an old mine that was widened to accommodate the hoard of military equipment.
>
> He left the cave and hiked to the top of the hill above to find where the shaft was. It was surrounded by three large white elongated stones and covered with a large flat rock. He returned to the entrance, where he stacked rock and dirt back into the entrance to reseal it. Later, he added more timber to make certain it stayed sealed.

A few years later, Ray Anderson and his partner drove to Springville Canyon in an attempt to locate the cave. After a few unsuccessful trips, they finally found the upright white stones on the hill above the cave. Since it was within sight of the road, it was decided that the two men would return the following night with all the necessary equipment to descend into the shaft.

The next night found the two men sneaking up the hill to the covered entrance. Once the flat rock was removed, they cast their lights into the shaft and could see down into the cave. Securing their rope to a large oak, they entered the shaft one behind the other, and down into the cave they went.

At the bottom, they were dumbfounded to see that what the woman had told them was true. Their heads went wild with ideas and speculations as to what to do next. It was decided that they would come back, find the front entrance, and move everything to a secret place somewhere in Provo.

Their plan fell apart when, after returning to the surface and climbing out of the shaft, bullets began to rain upon them from a cabin below and across the road. Ducking for cover, neither man had any idea why they were being shot at, but they knew someone had observed them going into the shaft and that, more than likely, it was their flashlights that gave them away.

The two men ran out of the area before one or both of them were killed by this unknown assailant. They decided it was too risky to return and carry out their plans.

Many years later, Anderson confided in a man from Ogden named Gary Falls about the discovery. Falls, knowing Anderson was getting up in years, convinced him to take him to Springville Canyon to show him where this place was; Anderson agreed.

I was invited to accompany Falls and Anderson to the site. Much had changed since Anderson and his partner had been there. The hillside had burned years ago, so where trees had grown, now tall grass and sagebrush covered the hillside.

It took several attempts but, at length, the white elongated rocks were found. However, they had been dragged, pushed, or slid part way down the hill, but Anderson knew full well these were the rocks that once surrounded the shaft. We tried to locate the shaft, but it was all in vain, for nothing was found. It had been covered by someone or something.

We tried to convince the property owners of the magnitude of this great discovery and that an excavation of the cave was in order. However, the owners wanted nothing to do with it and even rejected a plan to take them to the site. They couldn't care less. So, there it sits—a very important piece of Utah history that may never see the light of day.

OUTLAW LOOT AT MOSIDA

Mosida was a small village located on the west shore of Utah Lake about thirteen miles north of Elberta. It derived its name from each of its three promoters: Moore, Simpson, and Davis. It was to be a grand community where folks could come, build a house, and live a life of luxury. However, the operators of the new town became debt-bound, and the would-be boomtown died almost as quickly as it had started. The town began to crumble and the lumber used to build with was eventually scavenged away. What was left turned out to be a good place for outlaws to hide out.

One newspaper account tells of a lone bandit who stole a gold payroll from the army and headed south down the west side of Lake Mountain to Chimney Rock Pass. With posses hot on his trail, not to mention soldiers, he turned into the pass and headed toward Mosida. Just before he came out of the pass, his horse fell and broke its leg. The outlaw ran with his loot toward the town but didn't get very far. A gunfight ensued once the posse and soldiers caught up with him and the outlaw was killed. As he lay dying, he whispered into the ear of a posse member that he could see the ruins of old Mosida from where the bandit hid the gold.[9]

Utah Lake inscriptions

Not far from where it is believed the outlaw buried his gold is an old mine, and up on the rocks above it are signs and symbols that some say are by the Knights of the Golden Circle.[10] Did the outlaw have time to carve these inscriptions? Or are they fake?

THE DREAM MINE

Of course, at the mere mention of Utah County, the Dream Mine is immediately brought up. Many folks have been involved in this mine (or treasure, as some think). This mine is supposed to be the site of an ancient Nephite mine.

Dozens of stories have been written concerning the Dream Mine. Is it fact or fiction? It's really anybody's guess. The story surrounding the mine is long and complicated. It begins with a man named John H. Koyle, a faithful member of The Church of Jesus Christ of Latter-day Saints, who owned a farm in Leland, near Spanish Fork.

Koyle sold much of his produce in the Utah mining towns of Mercur, Ophir, and Tintic. One night, he had a dream of a rich deposit of ore located in the Tintic Mountains. He saw the exact location of the ore and where it lay in the mountain. He later related the dream to his LDS bishop, who told him that it meant nothing and to forget he had ever dreamed it. He was warned that trying to find riches through spiritualism was a tool of the devil and not to pursue it.

The farmer tried to forget his dream, until he heard that a rich ore deposit was found exactly where he had seen it. Koyle saw the discovery as a confirmation that his vision had been inspired.

Years later, on August 27, 1894, a heavenly messenger visited Koyle at his bedside and told him that he had been called to a special work. The messenger showed him a mountain east of Salem City and informed him that a rich Nephite mine had once been located at the site, thousands of years earlier.

According to Koyle, the messenger then began to lead him through the mountain along a worked-out mineshaft, all the while explaining various formations and tunnels that appeared along the excavation. "At an undetermined depth, Koyle was shown a tunnel he was to dig which would penetrate the mountain from an undisclosed point on the surface and intersect the excavation through which he had just passed."[11]

Koyle recalled later that he had "plainly seen mine cars carrying rich loads of ore"[12] in the mineshaft. He also observed five tunnels leading to rich deposits of ore. The messenger instructed him to name these tunnels the Five Fingers and informed him that the first ore shipped from the mine would come from the number four tunnel. Koyle was also told that this shipment of ore would set off the news of a great gold strike at the mine and would be the means of bringing much-needed relief to the people of this area.

The messenger returned to Koyle for two more nights, each night repeating the same instructions. Finally, on the third night, the messenger told Koyle that a neighbor (who had been digging a water well without success) would find water on the morrow at noon, and this would be a sign to Koyle that he should begin work on the mine.[13]

The next day, Koyle told his wife, Emily, about the dreams he'd been having. She didn't want to seem too skeptical, but she could hardly believe such a fantastic tale. He told her to watch the neighbor who was digging a well to see if they struck water at noon and then left for his daily labor in the fields. Emily was doubtful of the whole story, so she went about her duties as usual until she heard shouting and yelling from outside. She ran to the window to see what was going on and saw that the neighbor had just struck water. Emily looked at the kitchen clock—it was exactly noon.[14]

When Koyle returned from the fields that evening, Emily was waiting for him. She was so excited about the well that all she could do was point to the neighbor's yard and smile.

In September 1894, Koyle and five of his friends went to the mountain he'd seen in vision and staked out seven mining claims in the area. They organized themselves into three shifts that would work around the clock, with two men per shift.[15]

No mining stock was sold during the first few years of work on the mine, but on March 4, 1909, the Koyle Mining Company was incorporated. "John H. Koyle was listed as president and director of the corporation, with J. P. Creer as vice president and director, and W. Jones Bowen as secretary and treasurer. These three, plus George Hales, B. F. Woodward, and John F. Beck, made up the board of directors"[16]

Koyle continued to receive visions, which revealed more and more details about the mine. The dreams gave instructions on where exactly they should cut the access tunnel that would lead to the Five Fingers, what kind of ore they would find where, and visions of unique rock formations they would encounter next, which helped to convince him and those who worked with him that the visions were true and of divine origin.

Stockholders were attracted to the mine from every walk of life; most believing that it had a divine purpose. The stock was seldom ever sold to a non-Latter-day Saint, for this was primarily and essentially an LDS project, saturated as it was with LDS ideology and religious objectives. Nonetheless, it was completely independent of the Church leadership itself.[17]

In time, Heber J. Grant, president of the LDS Church at the time, and apostle James E. Talmage began to speak out against the mine and it was eventually closed.[18]

Today, the Dream Mine stands silent against the mountain. Little work goes on there. Some die-hard mining enthusiasts are still waiting for the promises made to Koyle that the mine would reopen and its riches cast out among the poor of humanity. While John H. Koyle never realized his dream, perhaps some of his offspring will.[19]

The rest of the story of the Dream Mine remains to be written. Was Koyle really visited by heavenly messengers? Is there really a vast treasure hidden deep within the bowels of Salem Mountain? Perhaps someday, we will find the answers.

THE TINTIC MINES

The old mines at Tintic have their stories to tell. Not only were they rich with gold and silver, but they contain hidden loot and stolen property taken by outlaws and wayward miners. Often, miners would find huge pockets of ore. One such pocket was discovered and the ore was loaded onto a wagon. It is said the load itself was worth $50,000 while some others went much higher.

High-grading (picking out the very best ore) was common in those days, and with that came thieving miners who would pocket high-grade pieces of gold and silver float (small pieces of gold and silver that float to the surface). Stealing a few dollars worth of ore was nothing compared to what was stolen right under the noses of the mine owners—a fifty-ton railroad car piled high with high-valued ore worth over a quarter of million dollars. How on earth did someone get away with a whole train car without being noticed? It seems there were a few people involved who were high up in the company. Six months later, the ore car was discovered empty in Mexico.[20]

SHERIFF BOYD AND THE MINE

In the 1920s, John D. Boyd Jr. served as the sheriff of Utah County and later as the Provo City Chief of Police. During his days as sheriff, he met his share of shady characters. Boyd had a reputation for being honest and tough. He was elected sheriff in 1920 and was supported by both Republican and Democratic parties. It was in the days of the Prohibition of liquor,[21] and Boyd soon became known as "the terror of the criminal and the bootlegger."[22]

George H. Brimhall, president of Brigham Young University, wrote the following note to Boyd in 1922:

> Sheriff J. D. Boyd,
>
> When you became a candidate for the office you now hold, some of your political opponents tried to convince me you were a "whiskey man," and I am convinced by your record that you are still a "whiskey man" of the anti-type.
>
> From my point of view, if all the offices of Utah County were filled with as much ability and honesty as that of Sheriff, there would be no need for an election, except as a vote of confidence.
>
> Sincerely yours for a continuation of good work,
>
> George H. Brimhall[23]

Boyd was so successful in upholding the laws that it is recorded from January 1921 to October of 1922, he and his deputies made 517 arrests for various crimes. Forty liquor stills were destroyed, 125 arrests were made of illegal possession of liquor, and $10,417 in fines were collected and turned into the county treasurer.[24]

Recorded in his journals, John remembers: "There were stills from Pelican Point on the other side of Utah Lake, right on around to the Provo side; up and down the hollows. One time a runner from Oregon had two eight-gallon kegs hid near Spanish Fork. I had myself a stakeout. When the fellers came driving in for their kegs, they found themselves looking into two six-shooters. They was real agreeable about loading the kegs into their car and driving me to Provo since I was in the back seat with a gun in their ribs!"

As tough as John Boyd and his deputies proved to be, John had a kinder side to him. Sheriff Boyd's regime at the county jail was characterized by a policy of kindness and fairness. Although he allowed none of his prisoners to be coddled, they did enjoy some definite advantages over most jails of the day. The Utah County Jail was one large room that was attached to the Boyd home in the back, and was located at First South and First East, in Provo.

John David Boyd Jr.
at ninety years old

The sheriff's wife, Hannah, was the jail's matron, and it was her duty to feed the prisoners. She would do this by preparing a meal, then sliding it into the cell through the windows of their home. John would often remark that perhaps some of the prisoners allowed themselves to be captured on purpose, just so they could enjoy Hannah's home-cooked meals. Many prisoners returned years later just to thank the Boyds' for the fair and humane treatment they received while in jail.[25]

During his time as sheriff, Boyd was called about a robbery in Springville. Upon arriving at the site, he was told the outlaw had headed up Springville Canyon on a horse. Sheriff Boyd acquired a horse at the mouth of the canyon and then headed up the canyon in search for the outlaw.

Boyd was an excellent tracker and followed the tracks of the bandit into Wadsworth Creek, a tributary to the Right Fork of Hobble Creek. Finding the outlaw's horse hidden in some oak brush, Boyd pulled his gun and began tracking the outlaw uphill through thick brush and trees.

As Boyd came through a thick stand of pine trees, he could see the mouth of what looked like a small cave, but upon further inspection, he found it to be an old partially caved-in mine. He knew he had the outlaw dead to rights so having positioned himself out of harm's way he called to the outlaw to come out and surrender. It took some convincing from Boyd before the outlaw figured he would either have to come out or stay where he was and most likely starve to death. He threw out his weapons and came out with his hands up. Boyd took him into custody.

The stolen money was inside the old mine; so, after securing the outlaw, Boyd went into the mine for the money and it was then he found writing carved into a headboard in the roof of the old mine that said, "Diciembre 1860." Looking around, he found good ore, both silver and gold. He was standing in an old Spanish mine.

Weeks later, Sheriff Boyd returned to the old mine site; however, he had a hard time finding it right away because of its great concealment. Boyd took samples out of the mine, then transplanted some wild rose bushes at the mouth of the mine, thinking they would grow up and eventually further conceal the mine, but would be easy to find by those rose bushes.

Sheriff Boyd loved to prospect. In his off-hours, he spent time in Utah's west desert county, hunting for mineral deposits. Whether or not he ever profited from the old Spanish mine in Wadsworth Creek is unknown either by friends or family.[26]

THE TREASURE OF THE DEAD TRAPPER

"A tunnel filled with gold ingots, some covering the body of a dead trapper, drew like a magnet on Johnny Rasmussen's mind for fifty years, to lead him from Old Mexico to the state of Utah in . . . 1913."[27]

Johnny had first heard about the treasure from his great-grandfather.

> In about 1825 . . . , Johnny's great-grandfather was with a trapping party in the Rocky Mountains area.

> It was getting along to early fall. Fur bearing animals were becoming scarce so the trapping party held a powwow in a meadow southerly from the south end of the salt sea. . . .
>
> It was decided to separate into smaller groups of three and spend about ten days to two weeks . . . explor[ing] the area for the next spring's trapping.
>
> Rasmussen and two companions were sent in a southerly direction following a large stream. For two and one half days, they explored branching streams but kept to the course of the starting point stream as a guide line.[28]

Following what we now know as the Jordan River, they came to a very narrow canyon through which the river flowed . Once on the other side of the narrow canyon, they beheld a beautiful valley with a large fresh water lake, which we know as Utah Lake today.

They saw a number of streams that flowed from the mountains to the east, which they explored for beavers, but found none. At the south end of the valley, they found a little stream that had evidence of beavers being there. They followed the stream toward the mountains to a hill they called Sugar Loaf that stood at the base of a large mountain. They were making their way toward the hill when they observed the camp of some local Indians.[29]

> Not knowing whether the Indians were friendly or not they skirted to the north of the camp and headed for the hills. They were exploring little side streams that fed the main creek when suddenly they saw a party of Indians coming from the south. They forced their horses higher up the hills and through the cedar trees. One of the Indians sent an arrow which lodged in the back of one of the trappers. He hung on doggedly to his saddle until they saw a badger come out of a large hole on the side of the mountain. One rider . . . found the hole to be large enough [to conceal them all].[30]

They pulled the wounded man through the brush and over the sharp rocks down into the hole. The Indians did not see where they went so they felt lucky to still be alive.

As evening approached, the temperature cooled, so the men collected dried sticks and brush that had been pulled down into the opening by animals and built a fire at the entrance in an attempt to stay warm. "One knotted cedar branch made for a good torch and was taken from the fire to light the cavern to its depths."[31] The men could see that their refuge was much more than a badger hole; it was the entrance to a fairly large cavern.

The light was taken deeper into the cavern, and before getting too far in, they happened upon a pile of gold ingots. Excited by this new discovery, "It was decided that the gold bars were too heavy for one man to carry without horses and the gold could not be continually hidden." If somehow someone found out about their cache, their lives would be in grave danger. "They came to the decision that the best thing would to be keep in mind the guide posts that would lead to their find until paper and pen could be had to draw a map and write a thorough description which would lead back to the sugar loaf gold."[32]

Shortly after this decision was made, they found their partner and friend had succumbed to his wounds and died. Johnny's great-grandfather and his remaining partner pulled their dead friend back to where the ingots were and then, one by one, covered his body with the gold ingots to keep wild animals from ravaging the body.[33]

Eventually the two remaining trappers escaped the marauding Indians and headed back to the Great Salt Lake to meet up with the main group. Johnny's great-grandfather was sure that some day, he could return to the hidden Sugar Loaf gold. But his years were consumed with marriage and making a living, and old age finally ended his lifelong dream of finding the treasure.

Before he died, he wrote down the most essential things that would guide him back to the gold ingots piled over the body of his dead trapping partner. He also described the treasure location as

> 1. Two and one half day's ride from the south end of the salt sea, and follow a river which [runs] northward from a large fresh water lake located in a beautiful valley.
>
> 2. Half a day's ride southerly from the eastern shore of a fresh water lake to a sugar loaf peak.
>
> 3. The sugar loaf peak is south easterly above an area with many springs that make a valley at the foot of the mountains and supply an Indian camp with water.
>
> 4. The gold tunnel is about three quarters of a mile south of the sugar loaf peak and high on the foothills.
>
> 5. The sugar loaf gold tunnel is below some rusty red ledges.[35]

In addition to his great-grandfather's writings, Johnny Rasmussen kept a map that his great-grandfather made of the hidden cave and used it when he came to Utah in search for it. The map was drawn on buckskin and Rasmussen kept it close and would not show it to anyone.[36]

After searching for a while, Johnny had to make friends and trust them to help him find the lost cave of gold. He eventually showed it to a young man in Springville, who exclaimed, "Johnny you have come to the right place. There is an old fellow south of here who has a dream mine. I think it's what you are looking for. He was shown nine rooms filled with gold bricks and ancient records in a dream and according to your record it is the exact location."[37]

"A few days later Johnny's story had reached the miner Bishop John M. Koyle, of Leland, Utah . . . , a little farming community just south of Spanish Fork, Utah."[38]

At this point, we pick up the narrative of the late author Benjamin Vern Bullock:

> Johnny was invited to the Koyle mine and there, in the cabin of the new dream mine[39] workings, told his story. After he finished, John Koyle, the owner of the mine, summed it all up and said, "Johnny, you are in the right spot. The way your great-grandfather traveled, exploring streams and heading south, it would have taken him about one and a half days to climb up the foothills from the easterly shore to where we are now. This canyon is the north side of the pyramid or sugar loaf mountain. Below you where Salem pond is now used to be springs. If you look to the south the valley closes so Johnny you are home." Johnny was a little confused and said his great-grandfather rode through cedar trees in the valley and on the foothill. Koyle explained to him that the pioneers or early settlers in the valley cut down the cedar trees for fuel and fence posts, denuding the area. Johnny was all but converted, then he remembered and asked Koyle if he had any rusty ledges. Koyle said they didn't but the tunnel was started in a ledge and they had an iron blowout high on the mountain. This didn't satisfy Johnny and he had to ponder over the location of the dream mine to consider whether it was the location for which he sought. It just came too easy to be brought to where someone else was seeking the same treasure.
>
> Lars Olson, the superintendent of the dream mine, sat in on a meeting. When Johnny asked about the rusty red ledges it rang a bell in Olson's memory. He had a neighbor in Provo, Utah where he lived who owned a little mine fifteen miles south from Koyle's mine. Ben Bullock, the owner, had some red ledges. The whole location of the two mines fit to a tee; the story in Johnny's description with the exception of the rusty ledges which Mr. Bullock's had and Koyle's didn't. Also on a rise

above the springs by Bullock's location were the camping grounds of old Chief Blackhawk. It may have been one of his warriors that put an arrow into the back of the trapper. When Olson returned home for the week, he contacted Bullock and told him Johnny's story. This led to locating Rasmussen and arranging a trip to Bullock's "Golden Relief Mine" just south of Spring Lake, Utah. This little lake had been made by building a dam to back up the waters of the springs which were below the Indian camp. On a Monday Bullock hitched his team to his three seated buggy and with Olson collected Johnny for the trip to the Bullock mine.

Bullock had two mines. One drift was in a little canyon below some rusty red ledges. The other one was along tunnel high on the mountain in yellow rock canyon known as the Syndicate Mine. As the party rode south out of Payson Mr. Bullock pointed to the mountain farther south ahead of them and said, "This is your sugar loaf peak. The springs are a little north and below it is where a little lake is now. Chief Blackhawk had his camp on the rise above the springs. About three quarters of a mile south is my little mine. Above it a couple of hundred feet are some rusty ledges. If you look to the south you will see that the valley closes and if you look to the right you will see Utah Lake. Your map fits the dream mine exactly except for the red ledges."

When the party arrived at the ranch below Bullock's "Golden Relief Mine," Johnny acted a little disgruntled. He looked at the valley and foothills and saw no cedar trees. Mr. Bullock told him the same story as Mr. Koyle had about fencing and burning the trees for fuel. . . . They started toward the mine. . . . Something which looked like an old mine dump was found behind some brush and below the red ledges.

All of the men got excited and with enthusiasm began excavating the dump level. They brought in teams, plows, scrapers and other equipment. They spent a week trying to find the lost tunnel entrance. Johnny was so excited he expressed his desire to . . . move his residence to the site. . . .

The next two days were spent in getting equipment collected for the treasure hunt. Thursday afternoon [was spent preparing the site for excavation].

The second day of excavating brought the most excitement of the project. A plow point hit a cedar tree stump about eight inches in diameter. . . . [It] was left as a happy reminder to Johnny that his great-grandfather had truly ridden through cedar trees. About eight feet further into the hill a pile of charcoal was found. This the men sacked up to use for black smith purposes. This charcoal find led to the story of finding an old slag dump several years before. . . . It was evident that

> smelting of ore had been carried on in the vicinity during ancient times. It is believed that this work may have been done even before the time of the Spaniards. It has also been learned since then that back in those "ancient" times, mining was done by building a fire in the face of the tunnel heating up the rock. Then, cold water was poured on the rock causing it to fracture which made it easier to remove. Perhaps this is the reason for the stockpile of charcoal.[40]

Two weeks of hard work proved fruitless; there was no evidence of the treasure being there. More digging took place at different points on the mountain, attempting to access the old tunnel. Cold weather set in and the mining slowed. Johnny still firmly believed that the large cedar tree stump held the clues to his great-grandfather's lost gold treasure. As winter approached, Johnny Rasmussen left Springville, never to be heard of again.[41] Of course, his famous map went with him, as well as any other information he may have had that he never dared share with anyone.

Sugarloaf Mountain stands as sentinel over Johnny's lost treasure cave. Even though many have tried to access the hidden treasure site, none have accomplished their goals. Maybe the right person, or persons, just haven't come along yet.

The ledges at Sugarloaf

The Sugarloaf above Spring Lake, Utah

NOTES

1. Rufus Wood Leigh, *Five Hundred Utah Place Names: Their Origin and Significance* (Deseret News Press: Salt Lake City, 1961), 101; italics in original.
2. Ibid., 102–3; italics in original.
3. See "Lost Indian Mine in Utah Valley," Randy Bradford, *The Treasures of Utah*, forum, accessed April 12, 2017, thetreasuresofutah.yuku.com/topic/2441/LOST-INDIAN-MINE-IN-UTAH-VALLEY-George-A-Thompson-1977#.WO5XnlMrKit.
4. Ibid.
5. Interviews conducted by Stephan Shaffer, the notes of which are in his possession. See "Lost Indian Mine in Utah Valley," Randy Bradford, *The Treasures of Utah*, forum, accessed April 12, 2017, thetreasuresofutah.yuku.com/topic/2441/LOST-INDIAN-MINE-IN-UTAH-VALLEY-George-A-Thompson-1977#.WO5XnlMrKit.
6. Thomas P. Terry, *United States Treasure Atlas*, (Specialty Publishing, 1985).
7. "Camp Floyd," Audrey M. Godfrey, *Utah History Encyclopedia* in "Camp Floyd" *Utah Division of State History*, accessed April 12, 2017, heritage.utah.gov/tag/camp-floyd.
8. Ray Anderson, interview with Stephan Shaffer.
9. Story told to the author.
10. For more information on the Knights of the Golden Circle (KGC) see *Wikipedia*, s.v. "Knights of the Golden Circle," last modified April 4, 2017, en.wikipedia.org/wiki/Knights_of_the_Golden_Circle.
11. James R. Christianson, "An Historical Study of the Koyle Relief Mine, 1894–1962" (1962) *All Theses and Dissertations*, Paper 4598, 13.
12. Ibid.
13. Ibid., 17.
14. Ibid.
15. Ibid., 18.
16. Ibid., 20.
17. Norman Pierce, *The Dream Mine Story* (1972), 19.
18. James R. Christianson, "An Historical Study of the Koyle Relief Mine, 1894–1962" (1962), *All Theses and Dissertations*, Paper 4598, 24.
19. "Apocalyptic Paydirt in Utah," Eric S. Peterson, *Medium*, accessed April 12, 2017, medium.com/alt-ledes/apocalyptic-paydirt-in-utah-3107f2c957c5.

20. Thomas P. Terry, *United States Treasure Atlas*, (Specialty Publishing, 1985).
21. Frank Curreri, "Sheriff's adventurous life inspires book on Wild West Lawman faced tragedy, Apaches with aplomb," *Deseret News*, October 25, 1999, www.deseretnews.com/article/724425/Sheriffs-adventurous-life-inspires-book-on-Wild-West.html.
22. Ida Boyd Reid (daughter of John D. Boyd Jr.), interview by Stephan Shaffer.
23. Note in possession of author's wife, Bonnie Jane Shaffer.
24. Ida Boyd Reid collection on the Boyd family, 1839, Special Collections, Harold B. Lee Library (Provo, Utah).
25. Personal notes of Bonnie Jane Shaffer.
26. Ida Boyd Reid (daughter of John D. Boyd Jr.), interview by Stephan Shaffer. See also Ida Boyd Reid collection on the Boyd family, 1839, Special Collections, Harold B. Lee Library (Provo, Utah).
27. Dale R. Bascom, *Following the Legends: A GPS Guide to Utah's Lost Mines and Hidden Treasures* (Springville, Utah: Cedar Fort, 2007), 7.
28. Ibid., 8.
29. Ibid.
30. Ibid.
31. Ibid.
32. Ibid.
33. Ibid.
34. Ibid., 7.
35. "The Sugar Loaf Golden Treasure," accessed April 12, 2017, slimsgold.com/sugarloaf.htm.
36. Dale R. Bascom, *Following the Legends: A GPS Guide to Utah's Lost Mines and Hidden Treasures* (Springville, Utah: Cedar Fort, 2007), 9.
37. Ibid.
38. The New Dream Mine was located up Water Canyon.
39. Dale R. Bascom, *Following the Legends: A GPS Guide to Utah's Lost Mines and Hidden Treasures* (Springville, Utah: Cedar Fort, 2007), 9–10.
40. Ibid., 11.

Wasatch County

"Wasatch County is an irregular area of high altitude—on the top of the Wasatch Mountains from which it is named. In the well-watered meadowlands of the Upper Provo [River], Etienne [Provost] held his rendezvous with trappers, traders, and Indian associates during the trapper era; and today this valley is the home of the main population of Wasatch County. Highway 40 traverses the county from the northwest to the southeast up Daniel's [Canyon]."[1]

TRUELOVE'S LOST GOLD

A man by the name of Truelove Manhart found a rich gold deposit that assayed at $50,000 to the ton. His find was somewhere near what he called Hayden Fork near Park City. Unfortunately, a forest fire swept the area and destroyed so many of his markers that he was never able to relocate his rich gold deposit.[2]

MYSTERY GOLD

In 1893, two prospectors searching a deep, abandoned mine shaft found the skeletal remains of a human and a rusted lunch bucket full of extremely rich gold ore. This man evidently fell into the shaft while returning to Park City after finding a rich gold deposit somewhere in the area. No gold was found in its natural state in the old mine.[3]

SPANISH GOLD IN HEBER VALLEY

During the 1850s, when Heber Valley was being settled, stories were often heard of Spanish and Mexican miners roaming the surrounding hills. At times, actual encounters were had between the settlers and the miners. Most of the time, the Mexicans or Spaniards stayed clear of anyone that wasn't part of their company. Local Indians told the settlers that the miners had been in their country for many years taking gold, silver, and slaves back to their homelands.

Many of the settlers didn't have the time or the resources to go out hunting for gold, but had to eke out a living by farming or trading. However, as time went by, people began to find the time to explore the mountains, for purposes other than for lumber and game; they went into the hills to find gold and silver. Discoveries of old Spanish spurs, swords, and even cannons made it all the easier to hunt for those lost mines and treasures that people heard so much about.

BOREN'S AND BETHERS'S TREASURE ROCK

In 1896, William Bethers and his partner Henry Boren discovered two ancient looking mines, about two miles north of Daniel's Canyon. The mines were driven in solid rock next to a large slide, making them hard to see because of the loose rock everywhere.

Bethers reported that the mines were expertly done and were large enough to get a donkey in and out. Not too far from the mines on the slope of a hill, Bethers discovered a large boulder that had what he described as "peculiar looking hieroglyphics cut into it."[4]

Not knowing how to translate ancient hieroglyphs, he engaged a man from Salt Lake City to accompany him to the site for the purpose of deciphering their meanings. However, the man could not give Bethers what he wanted, so Bethers was left without ever knowing what the strange characters meant. Both Bethers and Boren believed the signs were a road map of sorts for miners returning to the canyon after a long absence so they could easily find the old mines.

The two men worked the old mines in between farming and tending to their individual families' needs. They opened one tunnel to a depth of seventy-five feet and the other to only twenty-five feet. Stories flew concerning the old mines, and people began spreading rumors that the two men had struck it rich. However, there was never any actual proof that they ever profited from their hard work at the mines. Eventually, Boren was

Is this Boren and Bethers's treasure rock?

called on a church mission, and Bethers moved away, leaving the mines to the forces of nature.[5]

OLD MAN WOOLSEY'S MINE

Jim Woolsey, from Benjamin, claimed to have discovered an old mine tunnel while herding sheep above Currant Creek Reservoir. Here is his story:

> James B. Woolsey was a sheepherder all of my life and had a prospector friend who was a dentist in Salt Lake City, Utah. The dentist had him bring samples to him from various areas where he herded sheep. If they looked good the dentist had them assayed. The dentist promised Jim if he ever found anything good he would stake it and give him a percentage.
>
> One particular year Jim took his herd of sheep above Currant Creek just east of Low Pass on the summit. He also made camp on the summit just east of Low Pass. Several days later he rode his horse down over the hill, north and east of camp. He hadn't been gone long when he spotted a mine tunnel. He got down from his horse and looked at the tunnel. At the entrance the props were sticking out of the ground. . . . [The rock] was a dark reddish brown in color and in the rock was a gold colored metal which he thought might be iron. He chipped some samples from around the tunnel to take to the dentist. During the time Jim was on the mountain he rode by the old mine tunnel several times. That fall when

he returned to Salt Lake City, Jim took the samples to the dentist's office only to find out he was out of town at the time. He left the samples at the dentist's office anyway. Within a couple of days Jim got a job offer in New Mexico and left Utah for a time.

When he returned the next year to visit the dentist he learned that he was in a mental institution. Jim found out that the dentist had had the rock assayed and that it ran $57,000.00/ton. He also discovered that the dentist had spent most of the winter and the spring hunting for Jim until his nervous collapse. Jim did not know whether or not the ore he had brought him was the cause of his condition, but with that in mind he never went back to the old mine.

Jim gave directions to find the mine and said one should have no trouble finding it. Here are the instructions:

Go up Current Creek. Near the head is a road that takes off to the right. Follow that road up the mountain and over the top. Go east through Low Pass and continue going east along the summit. East of Low Pass you can find the camp where Jim stayed. You will know it is the place because all the trees have Woolsey carved in them. Just south and down the hill is where you can find some old dipping vats that were used to dip the sheep in. After finding Jim's camp go a little northeast from there, down over the hill somewhere around ¼ mile. The tunnel is in the timber and from the tunnel you can see the west fork of the Duchesne River.[6]

Two men, Jean Blitch and Cecil Dalton, asked Woolsey if he would accompany them up to the mine area. He said he would, so a time and date were set. The day chosen was in late October and it had snowed, so going was slow. Not far from Low Pass, the trio could go no further, so they had to turn around and return home. Blitch and Dalton never found the old mine, but they did find many of the clues Woolsey told them about. Woolsey died later that same year, but not before giving more information to a few special friends, who still hunt for that elusive mine on the west fork of the Duchesne.

Before his untimely death, Woolsey told his nephew Lars, "Go east of Low Pass, to where you will find my name carved on a large aspen tree. The mine is only a short way down the mountain from where I camped, in some thick brush and trees. You can see the mine from the Bobby Duke Trail."[7]

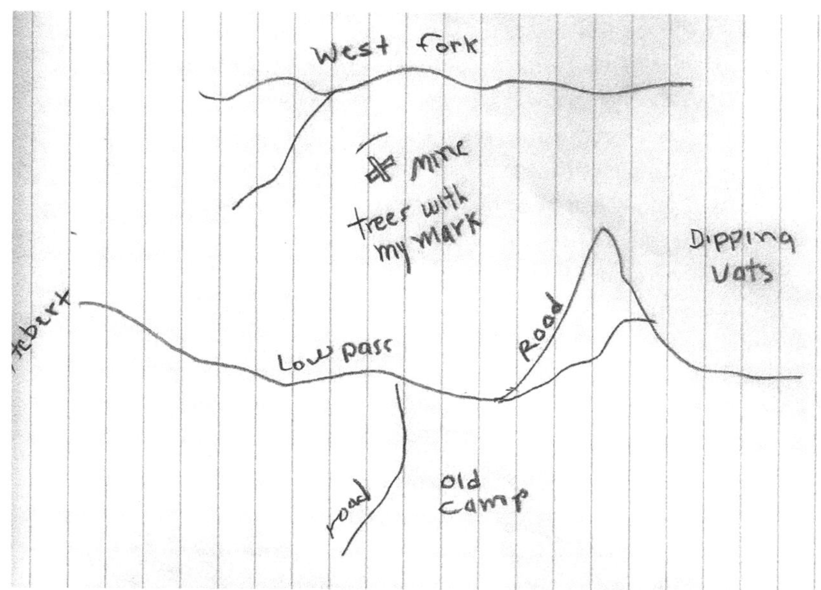

Lars Woolsey's map of the mine area (probably unreliable)

THE FULLER-THOMPSON EXPEDITION

The late George Thompson, author of several books and a prospector in his own right, met a man by the name of Captain Darwin Fuller from Santa Paula, California, who had acquired a small research grant from the University of New Mexico to trace some of the Old Spanish Trails within the state.

Thompson had read about the research Fuller and his team were engaged in. Thinking he could use them to his advantage, he contacted Fuller and convinced him to come to Utah with his team to help him trace Spanish trails and perhaps, in the course of that action, find a mine or two to offset their expenses.

Fuller and his team eventually found their way to Vernal, where they met and befriended George Thompson. The first trail Thompson wanted Fuller to investigate was the San Pedro Trail, mentioned in several old Spanish manuscripts and maps. This took the Fuller expedition across the Uinta Mountains to the high mesas above the Duchesne River.

Fuller recorded in his journal about the experience:

> In 1992, we followed the trail from N. W. from Kidney Lake to the head waters of the Duchesne where south of there on the west ridge we

located the three silver mines that are shown on the 1850s Renaldo map (Damian silver Mines) and the trails west to the ridge where the mine trails joined another major trek south of Trail Hollow. Later on, on the west ridge we located a fine map scribed on a rock that shows some of the north trails and mine sites north on the high meseta [plateau]. Several spots along the ridge have rock retainer walls that held the trail wide enough for the use of cartas [letters]. Just south of Trail Hollow, a bow hunter found a five-pound gold bar that George purchased. The irregular ingot had a lot of Spanish signs on it including the old marks sign that indicated the weight of 10 marks={5 lbs. plus or minus}. There were a lot of tree signs near the trails. We gave the three-layer 70 mm prints to George for his small support of our work. There are several area up there where infra red is of no use because of the dense growth, we sometimes get images among the trees that show outcrops but ground follow up is required to verify each one and many times the image may be iron or some black reflecting heat on rock. Our system can without ground work identify ore bodies in open country and the experts who take the blown up prints to special spectral reading boards have located some lucrative ore bodies prior to ground work. Soap Stone is wild with mineral but getting to it is going to be the test. Many "holes" in the area are barren with no mineral present. The problem we are having here in Utah is that we are being dogged on a regular basis by would-be thieves. I sleep with my gun at my side. One of our members Phil was accosted by two men who threatened him with death if he didn't reveal what we had found. Phil being a rather large man did not back down and hit one of the would-be assailants, which caused the two of them to run away. It makes for an uneasy time for us all.

George says he knows a retired Forest Service Ranger that claims to have found a small cave-like opening in a cliff which is about eight feet off the ground. He climbs into it with a ladder, which he conceals somewhere close by. Inside, there is a large room with a wall of white sugar quartz having a lot of fine gold in it. He told one son that the can fill a five gallon can in several minutes. From where he parks his horse trailer, he can go to there and back in one and a half hours, or about thirty minutes each way. The north end of Lightning Ridge, or somewhere very close by. The gold which Binggeli found in the shaft was also white sugar quartz, so soft he could break it loose with his hands. Both places are in the same area, so I can't help but think they are connected. What we need is an instrument that can detect an ore body which is probably close to the surface.[8]

As it turned out, Thompson and Fuller came to odds over a property that Thompson figured was his, even though Fuller was the one that actually made the discovery. Nothing was ever done with the discovery, so it was left to the winds of nature. Both Thompson and Fuller died before realizing anything from it.

NOTES

1. Rufus Wood Leigh, *Five Hundred Utah Place Names: Their Origin and Significance* (Deseret News Press: Salt Lake City, 1961), 105.
2. Robert F. Marx, *Buried Treasure You Can Find* (Dallas Texas: Ram Publishing, 1993), 346.
3. "Davis County," Floyd Mann, *The Treasures of Utah*, forum, accessed April 10, 2017, thetreasuresofutah.yuku.com/topic/971/DAVIS-COUNTY#.WOwImVMrKis.
4. "Old News Article from the Wasathch Wave," Randy Bradford, *The Treasures of Utah,* forum, accessed April 12, 2017, .thetreasuresofutah.yuku.com/topic/2288/Old-News-Article-from-The-Wasathch-Wave#.WO6Ll1MrKis
5. See "Old News Article from the Wasathch Wave," Randy Bradford, *The Treasures of Utah,* forum, accessed April 12, 2017, .thetreasuresofutah.yuku.com/topic/2288/Old-News-Article-from-The-Wasathch-Wave#.WO6Ll1MrKis
6. "James Warner Woolsey 'A Lucky Striker' 1826–1885," Gerry Shepard and Wilford Whitaker, *Ancestry*, accessed April 12, 2017, freepages.genealogy.rootsweb.ancestry.com/~woolsey/resources/descends/woolgen/wljameswarner1826_1885.html.
7. "Resurrecting the Lost Woolsey Mine," Randy Bradford, *The Treasures of Utah*, forum, accessed April 12, 2017, thetreasuresofutah.yuku.com/topic/1327/Resurecting-the-Lost-Woolsey-Mine?page=2#.WO6kKFMrKis.
8. George Thompson, interview by Stephan Shaffer.

Washington County

> Washington County, organized in 1852, encompasses the southwestern corner of Utah. . . . The altitude of arable and habitable parts is about 3,000 feet, the exposure is mostly southern, which result in a warm, equable climate, permitting the cultivation of cotton, sugar cane, grapes, and other fruits in pioneer days. In consequence of these crops the county early acquired a nickname: Utah's "Dixie." . . . Washington County was named to honor President George Washington; the name was selected when the Legislature was considering providing a block of Utah marble to be placed in the Washington Monument.[1]

Today, it is a bustling metropolis with important industry and caters to hordes of tourists and snowbirds.

Like many of Utah's counties, Washington County has had its fair share of historical events. The father of the county (the towns of St. George and Santa Clara, in particular), is none other than "The Peacemaker," Jacob Hamblin.[2] Sent by Brigham Young to make peace with the local Indians and to start a settlement, Hamblin started on a lifelong mission to the Indians of the Southwest.[3] Life was hard and, at times, just plain brutal to both Indians and whites alike. You had to be tough to live and to make a living in this inhospitable section of the state.

However, before all the glamour and luxury that we now behold, there was a time when things were not so rosy. Outlaws and highwaymen were the norm, and so were marauding Paiute, Navajo, and Mohave Indians. Gold miners and prospectors prowled the areas around the Santa Clara River and along the Old Spanish Trail, looking for mines, loot, or whatever they could find to sustains themselves.

LEAD AND SILVER AT THE SANTA CLARA RIVER

While en route to the California gold fields in 1852, a member of a wagon train, Jim Houdon, kept an ore sample he had picked up near the Old Spanish Trail not far from the Santa Clara River, about five miles north of the Shivwits Camp that later proved to be rich in lead and silver. After an unsuccessful prospecting venture in California, he returned to Utah, but abandoned his work, after rainstorms altered the course of the river. His discovery has never been located.[4]

PEGLEG'S SILVER LEDGE

"Pegleg Smith first discovered an extremely rich silver ledge in 1828, while traveling to California on the old Spanish Trail, near the mouth of Mogotsu Creek. . . . The site was close to the Santa Clara River's edge and at the foot of a red sandstone ledge. When he returned 25 years later to work the nearly pure silver deposit, he was unable to find it."[5]

WOOLLEY'S LOST COINS

> The firm of Woolley, Lund, and Judd was the St. George branch of the Mormon-owned ZCMI [Zion's Cooperative Mercantile Institution] store chain. It dealt in produce, livestock, mining equipment, etc., throughout Utah and Nevada, and marketed Dixie-grown cotton in California. As a convenience to its California customers, much of the firm's business was conducted through a bank in San Bernardino.
>
> In 1869 Frank Woolley made one of his regular business trips to San Bernardino over the Salt Lake Road. While there, he withdrew $20,000 in gold coins from his account and prepared to leave for Utah with a passing wagon train. He was delayed by business, however, and missed the wagon party, telling them that he would catch up with them later.
>
> Woolley hid two bundles of coins in five-gallon water barrels loaded on a pack horse. He left San Bernardino, crossed Cajon Pass and started out across the Mojave Desert. He was never seen alive again.
>
> When members of the wagon train reached St. George, they were surprised to find that Woolley wasn't already there, and thought that he had perhaps taken a shortcut and passed them on the trail. A search party found where he had camped at Resting Springs, near today's Tecopa, California, and followed his tracks eastward to where a group of Indians began following him.
>
> Not far from the Mormon settlements along the Muddy River, the search party found the spot where Woolley's animals had been taken by

the Indians. His body was found in a ravine, stripped and ravaged by coyotes, surrounded by empty cartridge cases. The trail of the Indians was followed, strewn with discarded items of Woolley's property, before being lost in Beaver Dam Wash. Various items of his stolen goods continued to show up over the next few months, but the water barrels were never found.

It is highly unlikely that the Indians kept the wooden kegs, as they were intimately acquainted with the local water holes and rarely carried water. It seems probable that the barrels were simply discarded in some nameless ravine. If that is the case, somewhere along the Old Spanish Trail in the vicinity of Shem lie sixty pounds of gold eagles, patiently waiting to be rediscovered.[6]

SPANISH SYMBOLS IN THE PINE VALLEY MOUNTAINS

Spanish gold hunters in the Pine Valley Mountains? Well, we know the Old Spanish Trail skirted the western side of the Pine Valley Mountains, but not on the east side, at least, not until a short while ago.

Several adventurous young men were hunting arrowheads in the foothills of the Pine Valley Mountains when one of the young men stumbled upon some inscriptions cut deep into a sandstone cliff face. The group spent the next several trips investigating the site and ultimately found several more inscriptions that resembled the first ones they discovered. Not knowing exactly what they found, they turned to a man (who wishes to remain anonymous) in the area that was somewhat knowledgeable in old Spanish symbology. He in turn contacted me to explore the area with him and one other gentleman.

Coded symbols

The area in question is off the main trails and seems to have been a central staging area for a group of miners that more than likely discovered a source of mineral wealth on the eastern slopes of the Pine Valley Mountains. The

Stephan Shaffer by the first symbols discovered

symbology indicated that coded messages were used to guide those with certain knowledge to the mineral site some five miles from the staging area.

The meanings of these symbols indicate a "direction of travel," but the code contradicts the meaning. The "turkey track" indicates the real direction instead of the "cross." A smaller symbol, like the one Todd Anderson is positioned by (on page 202), indicates a "small opening."

The symbols in this area can confuse a person not familiar with the complex system of symbols used by select Spanish explorers. Symbol association is the most important thing when trying to figure out what the author(s) intended. If a searcher has little experience or firsthand knowledge, they are at a distinct disadvantage. The Spanish would hide their best mines or mineral sites and would provide only one way to find them: to *know* the meanings and complexity of the signs and symbols left behind. Symbology takes years of work, both being out in the field and doing literary research.

Other symbols in this same area seem to mimic those shown. One other sign that seems odd, are the "footprints" that may or may not be authentic, but they give rise to another time when men wandered in this area.

Todd Anderson pointing to the "opening" symbol

Perhaps the Spanish had information that allowed ancient men to glean mineral wealth from the mountains to the east. Like most Spanish expeditions, they did everything possible to locate and plunder what they could by any means and no matter how many lives it might have cost.

Glen Cottam by the coded sign

One of two footprints in sandstone

NOTES

1. Rufus Wood Leigh, *Five Hundred Utah Place Names: Their Origin and Significance* (Deseret News Press: Salt Lake City, 1961), 106.
2. "A Brief History of Jacob Hamblin," *Utah's Dixie*, accessed April 12, 2017, www.utahsdixie.com/jacob_hamblin.html.
3. Marlene Bateman Sullivan, "'Friend and Brother': Jacob Hamblin, Man of Peace," *Ensign*, October 1984.
4. See Eugene L. Conrotto, *Lost Gold and Silver Mines of the Southwest* (Mineola, New York: Dover Publications, 1991), 121.
5. "Pegleg Smith," Floyd Mann, *The Treasures of Utah*, forum, accessed April 12, 2017, thetreasuresofutah.yuku.com/topic/952#.WO6t61MrKis.
6. "ZCMI's Lost Gold," *Old Spanish Trail—2005*, accessed April 13, 2017, www.quehoposse.org/index.php/plaques/15-old-spanish-trail.

Wayne County

> Wayne County is roughly a long parallelogram extending from Piute County on the west to Green River on the east; Sevier and Emery counties are on the north; Garfield on the south. Its principal physiographic features, from the west to east, are: the Awapa Plateau; northern part of high Aquarius Plateau; to its north is Thousand Lake Mountain; Frémont River flows southerly to near Bicknell, then, bisecting the high plateaus, runs easterly to Hanksville, thence southeasterly to the Colorado; east of the plateaus are Capitol Reef and Water Pocket Fold; beyond, the terrain is lower to Green River which moves through Still Water [Canyon]. Wayne County, organized in 1892, was named for the son of William E. Robinson, a member of the Legislature.[1]

Wayne County had its rough-and-tumble crowd that is well-known throughout Utah. The Wild Bunch is at the head of that list.[2]

ROBBER'S ROOST

The Robber's Roost was an outlaw hideout located somewhere in southeastern Utah. The hideout is famous for being used by Butch Cassidy and his Wild Bunch, but Old West lawmen never discovered its precise location, and at least three sites have been called "Robber's Roost." One of those three is in Wayne County.

"The hideout was considered ideal because of the rough terrain. It was easily defended, difficult to navigate into without detection."[3]

The roots of the Wild Bunch can be traced here. Elzy Lay and Butch Cassidy formed the gang while hiding out together. They made contacts

Josie Bassett

in nearby towns including outlaw sisters Ann and Josie Bassett, who owned a ranch.

"There were only five women known to have ever been allowed inside Robber's Roost: Ann and Josie Bassett, the Sundance Kid's girlfriend Etta Place, one of Elzy Lay's girlfriends Maude Davis, and gang member Laura Bullion."[4]

ORE MINING IN WAYNE COUNTY

Wayne County is known for its many mines, both open pit and underground projects. Uranium is most likely the number one mineral being mined in the county. Copper, vanadium, lead, barium, gypsum, and mercury are also mined there. However, there are gemstones that can be found throughout the county that may entice the reader to gather up his family and head to Wayne County in search for agate, jasper, chert, petrified wood, and selenite. However, make certain you have permission to search on private property or mining claims.[5]

FACTORY BUTTE CACHE

Wayne County is also home to Factory Butte, which has a long history of mining and ATV riding. There was a man by the name of Apache Jim Russell who cut out an ancient map that was inscribed in stone at the butte. He took the slab map to Sanpete County, where it was secreted in a cave with other artifacts taken from areas all over the West.[6]

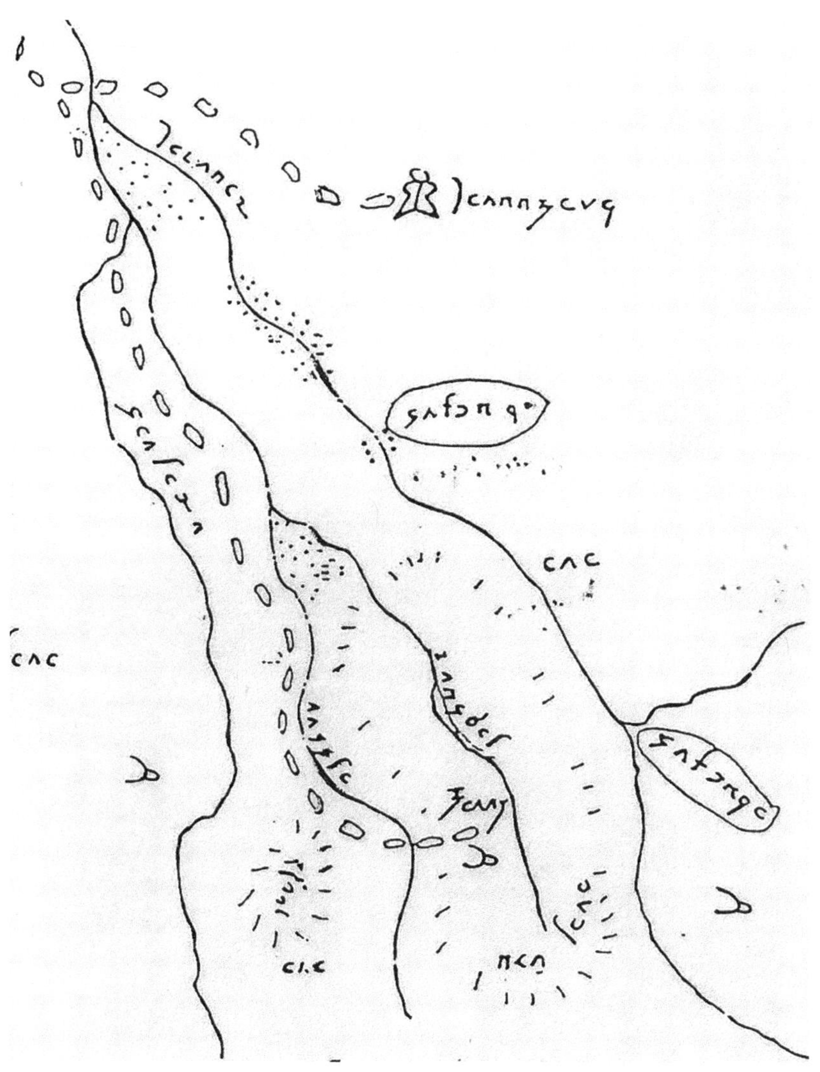

Apache Jim map

NOTES

1. Rufus Wood Leigh, *Five Hundred Utah Place Names: Their Origin and Significance* (Deseret News Press: Salt Lake City, 1961), 106.
2. "Butch Cassidy and Hideout Country," accessed April 13, 2017, redriverranch.com/2013/04/15/butch-cassidy-and-hideout-country/.
3. *Wikipedia*, s.v. "Robbers Roost," last modified April 2, 2017, en.wikipedia.org/wiki/Robbers_Roost.
4. Ibid.
5. See "Wayne County, Utah Deposits," *The Diggings*, accessed April 12, 2017, thediggings.com/usa/utah/wayne-ut055/mines; James R. Wilson, *A Collector's Guide to Rock, Mineral, & Fossil Localities of Utah* (Utah Geological Survey, 1995), 133; and Arthur Chamberlain, ed., "Mineralology," *The Guide to Nature* 2, no. 1 (April 1909), 211.
6. Note from Apache Jim Russell given to the late Earl J. Brewer of Moroni, Utah.

Weber County

"Weber County, one of the original eight, embraces the drainage of Ogden River, a magnificent section of the Wasatch Front, and the plain extending westward to the Great Salt Lake through which the lower course of Weber River, after being joined by the Ogden River, meanders to debouch into the lake. The county was named from the Weber River."[1]

The Weber River is the "second largest affluent of Great Salt Lake. . . . Weber [Canyon] together with its northeast fork, Echo, is the main gateway into Utah, having the lowest altitude through the Wasatch; it early became the route of explorers, trappers, immigrants to the Far West, Union Pacific Railroad in 1867, and modern Highway 30s. In 1825 a detachment of [General William H.] Ashley's men headed by John H. Weber explored and trapped down the river's narrow valley. Weber was one of four of William Sublette's men who circumnavigated Great Salt Lake in 1826. In the winter of 1828–29, John H. Weber was killed by Indians near the river. Sublette, factor of the Rocky Mountain Fur Company, named this important stream Weber River for his fallen man."[2]

FREMONT'S EXPEDITIONS

"[In 1844], [John C.] Fremont reached the [Great Salt Lake] . . . only one small part of a much wider journey of discovery and mapping. . . . Fremont began honing his skills as an explorer and mapmaker in his early twenties. . . . Fremont reached the Great Salt Lake during his second expedition. His 14 months of western rambling . . . resulted in the first comprehensive map of the Great Basin. . . . Fremont's maps became indispensable guides to thousands of overland immigrants heading westward to begin new lives."[3]

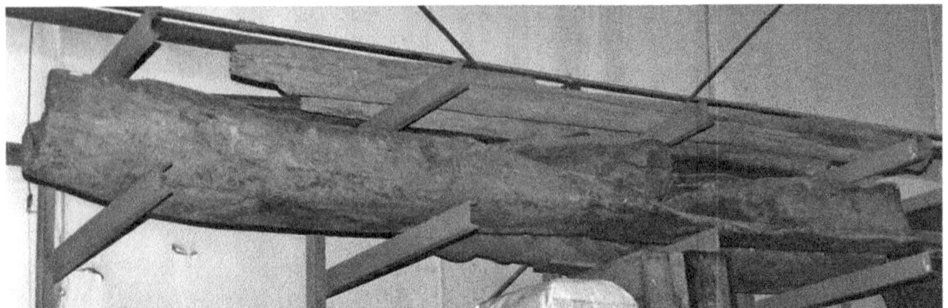

Dugout canoe used by John C. Fremont on the Great Salt Lake

NOTES

1. Rufus Wood Leigh, *Five Hundred Utah Place Names: Their Origin and Significance* (Deseret News Press: Salt Lake City, 1961), 106–7.
2. Ibid.
3. "John C. Fremont reaches the Great Salt Lake," *History.com*, accessed April 13, 2017, www.history.com/this-day-in-history/john-c-fremont-reaches-the-great-salt-lake.

Afterword

Within these pages, I have given little clues and stories that might be beneficial to the reader, whether hunting outlaw loot or just having fun chasing stories. However, the hunt is sometimes better than the actual find! So, have fun, and if you find something worthwhile, don't forget to write.

SPANISH MAPS AND MORE

Spanish expedition signs

Spanish expedition map

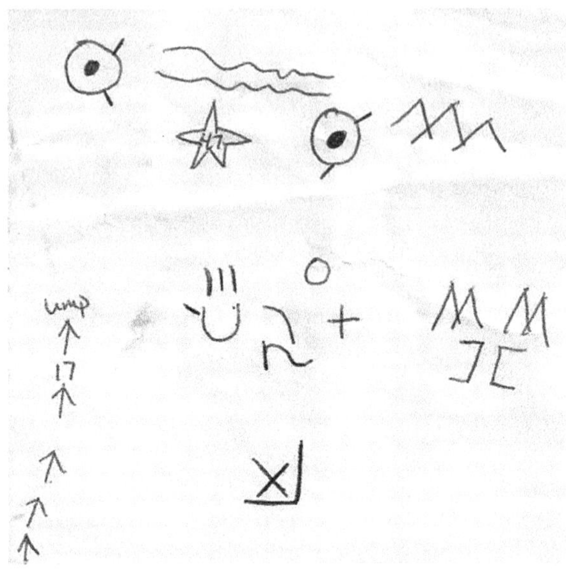

Ute map of Lake Fork and Upper Rock Creek, Utah

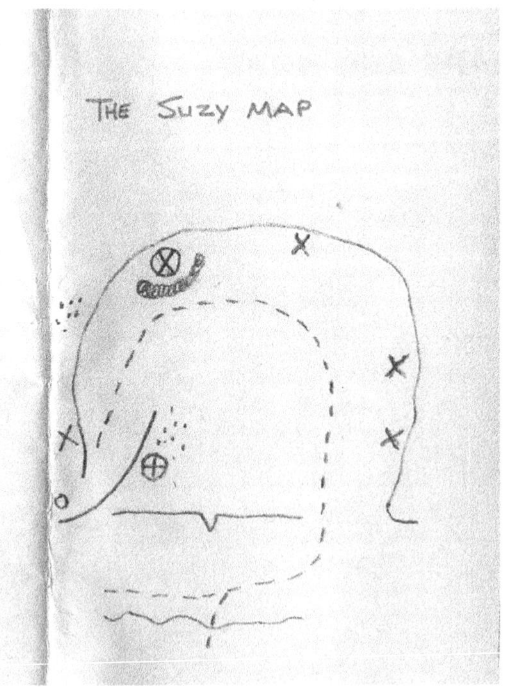

Ute map of Lake Fork and
Upper Rock Creek, Utah

Stephan Shaffer standing beside
"Map Tree" in the Uinta Mountains

Spanish map in Red Canyon, south of I-70

Sanpete County area map drawn by Johnny Brewer, circa 1975

Inscription in Little Egypt area south of Hanksville, Utah; possible caches hidden nearby

About the Author

As a Utah native, Stephan "Steve" B. Shaffer has spent most of his life researching Utah's ancient and not so ancient past. Beginning in 1964, Steve found that going into the High Uinta Mountains was a gift granted to him by a higher authority. He was always on the lookout for the unusual, the out-of-place things that would jump out at him. He traveled extensively throughout Utah, learning everything he could about past outlaws, Spanish miners, hidden treasures, buried outlaw loot, and more. Steve kept journals of his expeditions and amassed a huge collection of photos of his excursions.

He met and befriended many people along the way who were anxious to share stories about lost treasures and hidden loot in all the different counties of the state. Having a family and a full-time job made it hard to get out to those faraway places Steve so desired to visit.

Beginning in the mid-1960s until the present, Steve has not stopped researching and cataloging his every step along the way. During this time, he has befriended and has become blood brother to members of both the Apache and Ute tribes. His association with his Ute brothers, and others of that nation, has been instrumental in his pursuits for true historical events that took place in eastern Utah.

His helpmeet has been his wonderful companion and wife, Bonnie Jane (Peay) Shaffer. She has been his staff and wind beneath his wings, always prompting him to continue his work. Steve and Bonnie have been married thirty-one years, and, together, they have ten children and thirty grandchildren.

Steve has authored seven books and is continually working on others. He has a bachelor's degree in archaeology and a master's degree in education. Steve founded the "We Nooch Society: Historical Research and Preservation." He has since retired and is spending more time in field research.

Besides his many interests, he also makes American Indian crafts to sell and, at times, just to give to friends and family.